On September 26, 1944, an infantry battalion lay trapped under heavy enemy fire with no artillery support, facing the grim prospect of complete annihilation. Their only hope lay with a mortar unit dug in nearby, also under heavy enemy fire. To make matters worse, a direct hit on an ammo dump left the unit out of ammunition except for two hundred unexploded mortar rounds in the back of a disabled truck in an open field visible to the enemy.

A German Waffen SS sniper, one of the most deadly accurate shooters in the world kept vigil, scanning the truck continually through a scope mounted on a Mauser K98. Back to the field after accepting a decoration for his sixth kill, the sniper lay concealed in the detritus of a destroyed farmhouse six hundred meters from the broken down vehicle. Observing movement, he spied an enemy soldier charging the truck from out of nowhere, watching as he dashed from one large crater to another, covering nearly forty yards at a stretch until reaching the rear of the truck where he began moving rounds to a nearby tractor vehicle.

The sergeant's name was Lawrence J. Heron. When the call came for volunteers, no one dared risk a sniper's bullet, so the fastest and ablest man in the battalion decided to go it alone. A born athlete, Heron grew up in Hopedale, Massachusetts, a town at a peak of prosperity and recently designated America's model company town. His athleticism led him to St. Mary's High School in neighboring Milford, Massachusetts where he distinguished himself as a local hero in football and baseball.

Scholarship offers from colleges and universities piled up at his door, along with opportunities to play for two major league baseball teams, but war stepped in to alter his destiny. The Army instead assigned Private Lawrence J. Heron to the 87th Chemical Weapons Battalion, embarking him on a perilous journey that led to the outskirts of Cherbourg, France during the Normandy Invasion. There he found himself caught up in the greatest battle the world has ever known, an unparalleled invasion aimed at freeing Europe from the repression of a conquering fanatic named Adolph Hitler.

Ahead for Heron lay an inconceivable ordeal no mere human could survive, a suicide mission that left him the most severely wounded soldier to return from World War II. Driven to the limits of pain past the point that any living human can endure, Heron defied all odds, kept alive by one burning desire, to reunite with the woman he loved, the person who would serve as his eyes for the next fifty extremely eventful and extraordinarily challenging years.

Beyond Recognition

Greg Page

✯ ✯✯✯✯

Beyond Recognition

A World War II Story of Strength,
Perseverance, and Survival

Lowell Books
Jacksonville, FL

Printed in the United States of America.

Beyond Recognition: a World War II story of Strength,
Perseverance, and Survival / Greg Page.

ISBN-13: 978-0-9760428-1-5
ISBN-10: 0-9760428-1-9

Printing Number
10 9 8 7 6 5 4 3 2 1

Greater love hath no one than this...

You will never find a stronger-willed person than the Burn Survivor, for they have suffered the worst injury to the human body anyone can endure and survived.

- Gary R. Graham

CONTENTS

For Sue

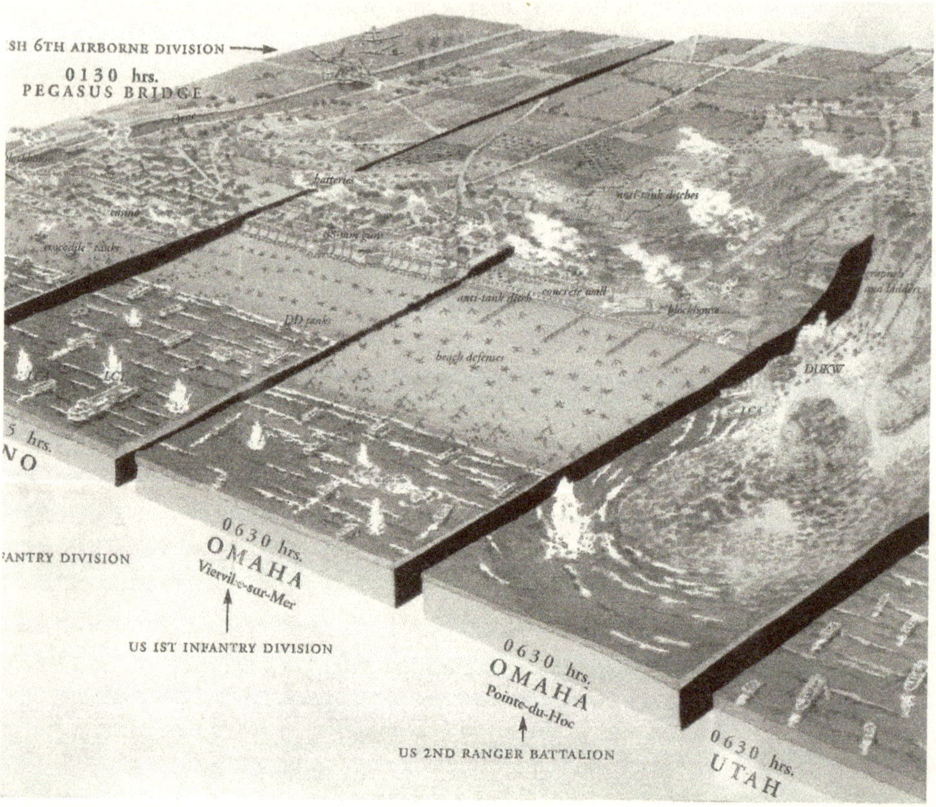

SH 6TH AIRBORNE DIVISION →

0130 hrs.
PEGASUS BRIDGE

batteries

casino

"crocodile" tanks

anti-tank ditch

Somme port

DD tanks

beach defenses

anti-tank ditch concrete wall

blockhouse

anti-tank ditches

tank landers

DUKW

LCI

5 hrs.

O

NO

ANTRY DIVISION

0630 hrs.
OMAHA
Vierville-sur-Mer

US 1ST INFANTRY DIVISION

0630 hrs.
OMAHA
Pointe-du-Hoc

US 2ND RANGER BATTALION

0630 hrs.
UTAH

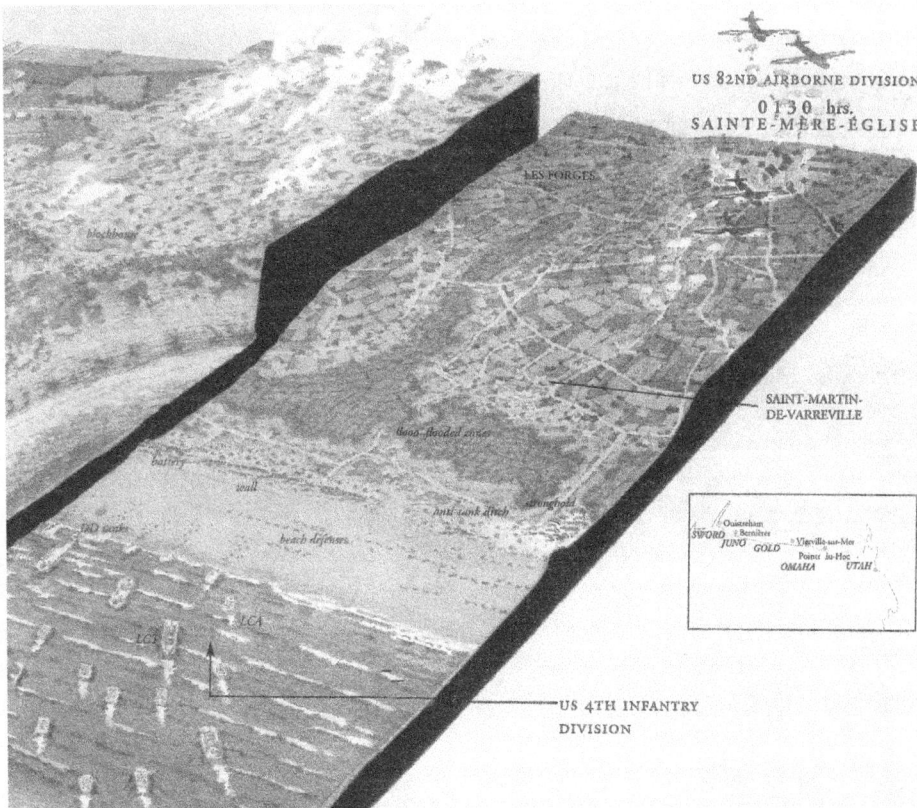

US 82ND AIRBORNE DIVISION
0130 hrs.
SAINTE-MÈRE-ÉGLISE

LES FORGES

blockhaus

SAINT-MARTIN-
DE-VARREVILLE

dune-flooded areas

battery

wall

anti-tank ditch   stronghold

beach defences

TD tank

LCT

LCA

Ouistreham
Bernières
SWORD
JUNO   GOLD   Vierville-sur-Mer
Pointe du Hoc
OMAHA   UTAH

US 4TH INFANTRY
DIVISION

Larry Heron scoring for St. Mary's, circa 1938
Courtesy of Larry Heron, Jr.

PREFACE

LIKE A SPIDER clinging to the edge of his web, waiting, patiently waiting, the Waffen SS sniper lay unmoving amid the ruins of a French farmhouse set on a knoll six hundred meters north of a two-and-a-half ton truck disabled by a land mine the preceding afternoon. Today marked his return to battle after receiving the distinctive grade-three Wehrmacht badge for recording his sixtieth kill. He would never wear the badge in combat knowing capture would lead to his instant and unceremonious execution. Hate boiled over on both sides for the men who issued one-way tickets to death, the well-known though unwritten rule of war that went with the territory.

Early on, the shooter learned two important lessons that kept him alive to this point. First, to perfect the use of camouflage to avoid falling under enemy crosshairs, and second to master the practice of extreme patience, waiting for as long as it might take for his target to come to him. From this lofty vantage point on the afternoon of June 26, 1944, he peered through his scope at a landscape commanding an excellent view of the back of the truck, where he knew the enemy might appear at any time to collect it or its contents.

Extreme patience rewarded him in less than an hour when the target suddenly popped into view, a single brave soldier charging from out of nowhere with a rifle carried loosely in one hand, hips swiveling, changing direction in an instant, and moving a bit too swiftly for the sniper to chance an accurate shot. He watched the target approach the stationary objective, drop the bed door, part the canvas draped across the rear then climb aboard to drag into view a large crate.

Centering his crosshairs on the soldier's back, he took up slack on the trigger, drew a deep breath and let half out, but just as his finger began to tense in preparation of ending yet another enemy life, the compulsion seized him to pause a moment longer, with the strong sense that a bit more patience could lead to higher rewards. The target would not go anywhere soon, and the sniper could take him out at a time of choosing. A shot now would kill one man, but in due course a single bullet could result in multiple kills, for he expected others to show up to collect the rounds from where this man was staking them.

For the moment at least, the killer of sixty men felt content laying the crosshairs on the middle of the shell the brave soldier lifted from the crate. The longer he delayed, the more brazen the enemy would become, and the closer it would bring him to his seventieth kill.

CHAPTER ONE

DEATH WISH

SHE LIES IN A hospital bed staring across the room at flowers her son delivered five days ago, on June 26, 2000, fifty-six years to the day her husband became World War II's most severely wounded soldier. She prevented the hospital staff from dumping the vase, perhaps because she identifies with the shriveled flowers and bent stems, so much so that she no longer can bear to look at herself in the bathroom mirror. It seems she's aged several long years these past few weeks alone. With labored breathing and lungs pumping like organ bellows, she presses her aching sides hoping to relieve the throbbing pain near her kidneys.

"Just the gall bladder acting up again," she told her son, Larry Heron, Jr., as he slipped her oxygen tank into the rear of his car for the one-mile drive from Hopedale, Massachusetts to the hospital in the neighboring town of Milford that morning. "Don't look so worried," she told him. "The doctor wants to keep me a few days for a thorough checkup, like last year." He recognized it as a cover but simply smiled and nodded knowingly, refraining from an open display of concern.

Early Saturday morning, her oldest daughter, Patty, entered the hospital room for a visit, joined five minutes later by her second oldest, Carol. "Nothing to worry about," Azelia told them. "Just don't have any energy lately."

Neither labored breathing nor skin turned whiter than the face of a Geisha in sharp contrast with scarlet cheeks struck her children as anything new. Their mother spent the past five years tethered to a plastic hose looped over her ears and under her nostrils then winding under foot to where it connected to an oxygen tank on the other end. Every year her children delivered her here for a checkup, so they felt no real sense of urgency or alarm.

Two weeks after Heron passed away an ambulance rushed her to the hospital hemorrhaging through her nose and mouth, gushing out three pints of precious blood before they stopped the flow. It took doctors two additional years to diagnose her condition as pulmonary fibrosis.

"What a shame," friends exclaimed. "She deserves far better, especially after what that poor woman has gone through."

When her youngest daughter, Debbie, visited the next morning, Azelia asked, "Have you been praying for me?"

"Every day I pray you'll get better."

"That's not what I mean. Tonight I want you to pray that I will go to sleep and never wake up."

"Mom!" Debbie protested. "Don't say such things." But she noticed a huge difference in her mother today. Comparing how weak and emaciated she now appeared to any time in years past, her body seemed even more shrunken and her skin sagged loosely under the arms. Embossed veins on the backs of her hands wore a dull gray pallor, appearing ready to burst through skin that had lost its natural color.

"It's all right, dear," her mother told her. "Every new day I wake up trapped in this body tied to an oxygen tank with nowhere to go and nothing to do except watch television, and each day I pray it will end. Don't you see? I pray God will let me sleep beside him once again."

Debbie witnessed her mother's belief in God grow infinitely stronger with the passing of each succeeding hardship throughout her lifetime. Long after she lost her Larry, the misery and suffering continued. Jesuits view suffering like Christ a necessary step toward reaching heaven, a normal part of life, and Azelia came to accept their theory that it prepares us for life in the hereafter. Whereas the chances of finding find happiness after death seemed fifty-fifty at best, she could rest assure she would never find it here on Earth, not in her present deteriorating condition. That night, Debbie could not bring herself to pray for her mother to die and instead cried herself to sleep, praying, "God, please help Mom find the peace and happiness she seeks."

On July 3rd, Larry's concern for his mother escalated. Her color matched the white of the bed sheets and she struggled to draw a breath, each more difficult than the last. Finally she told him, "Don't worry, son. I'm fine." Adjusting the headrest so that breathing came easier, she looked him in the eye and asked, "They've forgotten him, haven't they?"

"No, Mom. People still remember Dad."

She sighed. "They tore the Honor Roll down because no one wanted to appropriate funds to maintain it, and last year they voted down a proposal to put up a new memorial in its place." She drew another deep breath. "Since Father Connors passed away they stopped holding the 9th Division annual memorial celebration."

Larry's brows furrowed as he leaned forward to take her hand.

"How will they remember him, what he did? The Lawrence J. Heron Chapter of the DAV moved to Medway," she continued, "and most of his comrades have passed on. Those still alive, the majority are in hospitals or nursing homes waiting to die. No one seems to care anymore, all they gave up, how they suffered. It's all been forgotten."

Her son tightened his grip on her hand and fought back tears while struggling to find words of comfort, but everything that came to mind would sound meaningless when spoken aloud.

She shifted her body and adjusted the plastic tubing under her nose then ushered in another breath. "We had such a short time together before, before..."

Larry Heron, Jr. traveled this road often with his mom but that did not ease his anguish. "I know, mom."

"He had so much promise, too much to have it stolen from him in the prime of life."

She too paid a toll for the bravery his father demonstrated on the battlefield, tolls paid in installments over the course of decades. Yet neither young Larry nor his sisters ever heard their mother complain, so her next question took him by surprise.

"Was it worth it?"

Larry did not allow the answer he truly believed reach his tongue. He had asked himself that same question many times and the answer seemed clear. Ungrateful nations abroad and the very nation his father and countless other young Americans like him gave so much to preserve its freedoms, the very nation that prospered most from their great sacrifices, seems long ago forgotten.

Larry cleared his throat and smiled over at her. "I never heard him complain."

"But people should at least remember what he did, all he gave up, how he suffered. Is that asking too much?"

"He told me he felt like the luckiest man in the world to have married you. You meant more to him than any part of him that he lost. You were the one person he cherished most, and because of you, he never felt bitter or lost his faith."

"He loved all his children and grandchildren as much as he did me, but never saw what one of you look like," she sighed. "Not a one. That God would take him without ever letting him see his beautiful children. It's...it's..." Her voice trailed off and she closed her eyes while a single tear took a familiar path down her cheek.

Larry gave her hand a light squeeze. "It's OK mom. I truly believe he's looking down from heaven and can see us right now. I often feel his presence."

"June 26, 1944, was our longest day, one that changed our lives forever.

THE ODOR of hospital food turns her stomach the moment the food wagon rolls off the elevator, and when it arrives at her door, she waves off everything except a container of strawberry Jell-O. At 7:30 p.m. the night nurse comes by to wrap up chores. Before departing, she asks what Azelia would like to watch on television.

"You can leave it turned off."

As the nurse heads for the door she calls over her shoulder, "Okay, sweetie, sleep tight."

Sleep is the last thing Azelia wants, certain that this time when she closes her eyes, sleep will last for eternity. She does not fear dying but before meeting her end would like to accomplish one last goal. She read many tales of how the dying see their lives pass before them in that final instant, but she does not choose to move on without first reviewing every detail of their lives, not just flashes in the waning seconds but a calm, careful review that spans a shared lifetime, leaving out as little as possible.

Wouldn't it be nice if she could use her remaining hours to look back at a life lived with the one man in this world she loved like no other for close to seventy years? Let him walk beside her once more before she departs from the world forever. At least her mental faculties remain sharp and she can still recall with great accuracy each precious moment they spent together, and unfortunately every horrific memory as well.

Temples pounding in this manner usually signals a full moon over-head, and looking out the window straight across from her bed she sees tonight is no exception. The leading edge of the moon appears as a mere sliver peeking in through the window on its left side, and on the north wall to her right, she sees it reflecting in a small mirror, big bright and round.

With eyes fixed on that tiny sliver of moon peeking in through the window, she allows her mind to drift back to an earlier time when life held such promise and meaning, back to the days of happiness where it all began.

CHAPTER TWO

THE BEGINNING

THE SUN SHONE bright in a stark blue sky over Hopedale, Massachusetts on July 10, 1926. On a low stone wall that ran a hundred feet along Hopedale Street, she saw him place one foot carefully before the other, his arms flailing from his sides like a tightrope walker fighting a stiff breeze. Rows of towering elms spread their branches above like ribs of opened umbrellas, a contiguous canopy dappling sunshine onto the newly-paved street below. As she moved along the sidewalk, she inhaled a cocktail of harmonious scents wafting from freshly cut lawns, flowering lilacs, and a variety of colorful flowers.

The sight of the little girl approaching caused Larry Heron to lose his concentration and nearly fall off the wall. Embarrassed she may have seen him jerking his bottom back and forth while moronically flailing his arms in circular motion to keep from falling, it came as a relief to see her attention directed to the gaggle of high school runners jogging past, their numbered shirts bouncing in unison with their feet striking the macadam. Gaining control of his body just as she directed her eyes toward him once more, he paused his balancing act, feigning a shift in interest towards the runners.

The little girl and her mother crossed the intersection of Hopedale with Adin Street, named for the town's founder, Adin Ballou and approached within thirty yards of Heron, closing fast. Adin was Hopedale's street of dreams, a half-mile stretch affording taunting glimpses of the

Draper estates past a screen of concrete walls, wrought iron gates, pampered trees and lofty evergreens.

The attractive brunette walking beside the girl wore a blue dress and appeared about the same age as his mother. He looked them over as they stopped to admire a shiny black Model-T putt-putt past, smooth and shiny as though hot off the factory floor. As they drew closer, the girl appeared more striking than at first sight, and as quickly as chewing gum loses its flavor, the challenges of funambulism lost all appeal. Leaping off the wall he moved to his mother's side twisting his body for a closer view and felt his pulse quicken as she passed within inches.

The world changed for the better from that moment on: eyes as blue and bottomless as the deep sea; hair to her shoulders reminding him of corn silk under a summer's sun; a blue ribbon in her hair and a short-sleeved white dress with soft blue trim that complemented the splash of cobalt in her eyes. With heart ricocheting in his chest, he felt compelled to say something intelligent to impress her, but only managed to stare.

Emma Heron looked down at the girl commenting, "I don't blame him. She's beautiful."

Heron felt his face flush.

"What's her name?"

"Azelia."

"Pretty. Like the flower?"

"Named after the flower but misspelled on her birth certificate A-Z-E-L-I-A, so we just left it that way."

"How old?"

Azelia released her mother's hand and approached the wall where she turned to lean back against it, trying not to show any interest in him or return his stare.

"Just turned six," Livia Noferi said.

Azelia cocked her head at Emma and a smile curled at both corners of her lips. "So, she'll start school in September?"

"Yes," Livia said politely, "and your boy?"

Emma patted Larry on the head with a proud smile while he made a face and twisted away as though her hand held fire.

"He turned six just last month."

Heron hated anyone talking about him and pretended he couldn't hear.

Azelia attempted to pull herself up onto the wall but it lacked a foothold so she turned away as a wasp buzzed her ear on its way to search for sugar while she waved a hand to shoo it away.

"Isn't that nice? They'll be in the same class." In a town the size of Hopedale, elementary school classes typically numbered around twenty-five children who would remain together through high school.

Azelia, age 11. *Courtesy of Larry Heron, Jr.*

The women exchanged names and pleasantries while Heron worked up the nerve to finally speak. Just then a noisy truck rolled past while the driver clanged a loud bell, shouting, "Fruits and vegetables. Come get your fresh foods!" Heron looked Azelia straight in the eye and finally found his nerve. The mothers weren't supposed to hear the words that gushed uncontrollably from him as the truck suddenly stopped and the background noise abated. They broke out laughing when they heard him speak aloud, "I like you." Heron's face turned crimson and he abruptly spun around and ran towards Adin Street without daring a glance back at them.

He would see no more of Azelia until school began in the fall when romance would take a back seat to shooting marbles, tossing baseball cards, and otherwise matching the capabilities of boys his own age. But fate had taken a first major turn and two paths collided that day in July, with future promises of all the passion, love, and emotion life offers. The years passed quickly during which both Heron and Azelia developed into exceptional athletes still managing to bring home good grades. When he became of age, Heron worked summer jobs part time and saw little of Azelia until the start of those teenage years.

Azelia Noferi bounded noisily down the stairs to the front hall of their house atop Mendon Hill that strongly resembled dozens of quaint white houses scattered through a section of town aptly called White City, homes built to house the families of Italians brought to America to work for the Drapers. Similarly, the company constructed brick duplexes just over the line in Milford to house Portuguese immigrants transported to America to work in their foundries.

"Mom, what time is it?" Azelia asked as she snatched a crisp Macintosh from a bowl set on a hall table brimming with shiny fresh apples.

"Ten minutes ago I told you it was four-twenty," Livia chuckled. "Now it's four-thirty. Better hurry if you want to catch him."

"Oh Mom. I don't know what you're talking about. I'm just going out front to practice field hockey."

"What about your homework?"

"Done," Azelia shouted as the screen door slammed shut behind her.

The maple trees outside their home formed a canopy of brilliant colors from alizarin crimson to cadmium red to yellow ochre and burnt sienna, names her artist brother Guido Noferi taught her.

She drew the collar of her windbreaker close around her neck as a stiff breeze swiped past the Noferi doorstep then buttoned her collar to shut out the fall chill air. This was her favorite time of year when her mother cooked stews and made pies, with the field hockey season in full swing.

In the kitchen, her mother smiled because Mendon Hill stood a good mile out of Larry Heron's way to his night job at Patrick's General Store on the corner of Hopedale and Mendon streets. The town did not grow many like Heron who could heft two hundred pound sacks of feed grain like light-weight beanbags. Pretending he walked this way for the exercise, he would amble up the hill about this time every night to find Azelia practicing field hockey in front of her house. The time she spent out there each night certainly did not prevent her becoming Hopedale high school's top scorer last year.

A gusting wind sent leaves dancing in the air like crystals in a snow globe as Azelia flipped the ball ahead then drove it across the lawn toward a net her father, David Noferi, stretched out for her between two large trees. Just then, Heron strolled into view. She did her best to ignore his presence as he turned to face her with hands buried in the pockets of a sports jacket emblazoned with a St. Mary's emblem.

"Hi Azelia. How's it going?"

"Oh. Hi Larry," she answered, barely noticing as he continued toward her. She dropped the ball and gave it a wicked smack then watched as it bounced along the ground and into the net. "Ready for Friday's game?" she casually tossed over a shoulder.

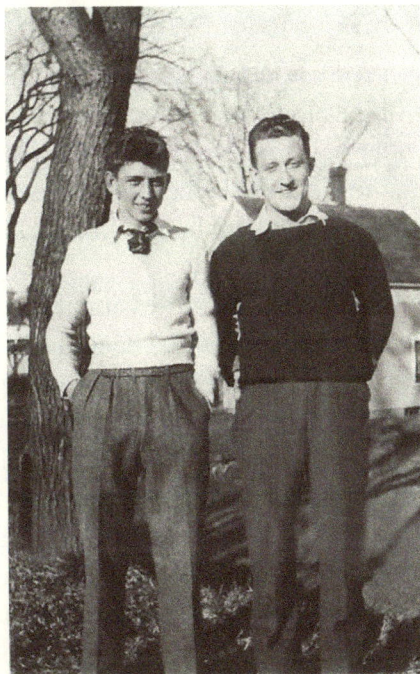

Azelia's brother Guido with Larry Heron, circa 1940.
Courtesy of Larry Heron, Jr.

Since that first day they met, she continued to worship him yet though that entire period they dated just once, at their first school dance held two months ago, just twelve days prior to her fourteenth birthday. They mostly held hands that night and Larry, too shy to ask for a kiss when the evening ended, walked her home around ten-thirty and sat talk-

ing to her on the porch steps for over an hour. Of course, it did not help
that her father left the porch light on or that they felt his eyes boring like
lasers through a front window. So when she opened the door for Heron
to leave, she did her part by turning to kiss him full on the lips. Then she
hastily said goodnight and dashed inside closing the door behind her.

Larry felt his face flush red and nearly lost his balance. With his
stomach flipping and the pressure of her lips lingering on his, he headed
for work, and during the minutes it took to get there, learned what people
meant when they talked about feeling butterflies in their stomachs. The
fragrance from her hair lingered and pangs of love gnawed at his heart.
Halfway down the hill, Heron let out a massive sigh and hummed softly
all the way to work.

JOHN HERON possessed a good Welsh name and had Irish-Catholic
blood flowing through his veins. For taking a bullet that just missed his
heart and left an exit hole in his back while fighting in the Boer War,
Queen Victoria pinned the last of five medals awarded him for service
in various campaigns, medals he kept buried in a drawer throughout his
entire life.

As a youth, John played semi-professional soccer, and when he
turned twenty-one, married an English girl named Emma Schofield. In as
many years they produced five children, none of whom made it to adulthood due to one disease or another raging through every city and borough
of Great Britain at the time. Whether from cholera typhoid, typhus, scarlet
fever, whooping cough, or pneumonia, no records exist to confirm, for
throughout the Industrial Revolution Great Britain suffered diseases many
and varied. Barely able to make ends meet, John seized an offer from The
American Print Company to relocate his family to Fall River, Massachusetts where he worked in the mill as a laborer. Five more children born to
the Herons in that American city would make it to adulthood including
John, Fred, Leonard, Ethel, and lastly Lawrence, in that order, and each
like their father, entered the world exceptional athletes.

American Print, the largest producer of printed cotton in the world,
sponsored a soccer team that played in a semi-professional league for
which John (son) and Fred participated just as soon as age would allow.
Upwards to 20,000 people would flock to Fall River on a given day just to
watch the gifted brothers make opposing soccer goalies wish they'd stayed
home. When wind of the athletic prowess of the Heron brothers reached

Draper Corporation scouts, the world's largest manufacturer of textile looms offered to relocate the family and provide jobs so the sons could play for their soccer team.

And thus the Heron family moved into a Draper-built home in Hopedale, Massachusetts, which they rented for $3.50 a week, for as long as at least one member of the household remained a Draper employee. John's two sons joined the Draper's semi-professional soccer team and thus began an era of Draper soccer dominance within the Blackstone Valley League, led predominately by the Heron brothers.

Endowed with super-genes and influenced by a sports-minded father and his locally famous brothers, it came as no surprise to the town of Hopedale when Larry Heron began setting new schoolboy baseball records and caught the eyes of local scouts watching him play sandlot football. Soon Hopedale's Sacred Heart Church approached Heron offering a full four-year tuition to St. Mary's High School in the neighboring town of Milford, since Hopedale possessed too small a population to field a football team of its own.

Heron wanted desperately to play football, and after begging his parents permission to transfer, accepted the opportunity to depart Hopedale High and transfer to St. Mary's.

Fr. Edward T. Connors, the new Athletic director at St. Bernard's in neighboring Fitchburg did not look forward to his team crossing swords with a Heron-led St. Mary's team at their homecoming in Milford. That year the loyal fans of St. Mary's had raised enough cash to replace old uniforms that literally hung in tatters from the player's backs. The players would appear in their new uniforms for the first time in the game with Fitchburg, and the Milford Daily News labeled them the "Rags to Riches Team," crediting the team's success in building a local following to team captain Larry Heron.

The Milford "Saints" reigned undefeated during this, Heron's first season, and Connors feared they would run ripshod over his St. Bernard's team, just as in the fall when they crushed his old alma mater, Northbridge High. It was at that game that Connors first viewed Heron as potentially one big thorn in his side.

But his team would be spared a similar fate as NHS because a week prior to the scheduled clash, fate intervened. In a game that week against the Angel Guardian's of Jamaica Plain, Heron fractured a rib while scoring his second touchdown. With their key back sidelined, the Saints went on to lose that and their next game against Hyde Park.

But Ed Connors differed from the average person in that none of this affected his righteous outlook. "I am truly sorry, for I have been looking forward to seeing Heron play against us this season," he told Father McCarron. "He's one of the most exciting backs in the league and it would have been a wonderful match-up."

Come game day, St. Bernard's upset St. Mary's handily. As both teams headed for their respective locker rooms, Connors approached Heron sitting dejected in his street clothes on a corner of a bench and offered a hand. "I did not feel the thrill of victory today," he said. "I had been looking forward to watching you play. Perhaps next year."

Heron rose to shake the hand extended to him. "Thanks, but this is my final year, Father."

"There's still a few games left in the season, and you will return to make up for lost time." Then Connors added with a grin, "Only thing is that when you do, I pity the opposition."

The priest left a strong impression on Heron for preferring to see him play even though it might cost St. Bernard's the game. Before departing, Connors added, "Perhaps we will meet again in some other place and time – if not on the playing field," a prophetic statement as it would turn out in later years.

Heron returned to the field the following Saturday, scoring two touchdowns against Hopkinton and leading St. Mary's to a 14-6 victory. Then came the classic Thanksgiving Day match up against Connor's Northbridge High School. On game day, Heron told Azelia, "I will score two touchdowns, just for you. That's a promise."

"Just to see you play will work for me," she answered.

THE SKY WAS a grainy gunmetal wash over Whitinsville's Picnic Point Field on Thursday, November 24, 1938 with an angry nor'easter blowing in throughout the past two days that left the field crusty and uneven, but the biting cold would not deter the four thousand shivering spectators who filed into the stands and pressed the edges of the field to watch the game get underway.

On St. Mary's first possession, Heron moved the ball thirty-yards to the fifty yard line. Two plays later, a trail of defenders lay sprawled in the turf as he crossed the goal line barely touched and still on his feet. In eight minutes, Heron had moved the ball seventy-five yards to put his team out front 6 to 0. That score would hold to the fourth quarter when Northbridge attempted a pair of desperation passes back-to-back, the second of which the Saints' quarterback intercepted and ran to his own 31 yard line. A pass play then moved the ball to midfield where Heron picked up three grueling yards, with half the Northbridge team riding on his back.

Another touchdown by Larry Heron. circa 1938.
Courtesy of Larry Heron, Jr.

With less than two minutes remaining in the game, it appeared his promise to Azelia would vanish like a puff of hot air. On the very next play Northbridge sacked the Saints' quarterback, who overthrew his receiver to get rid of the ball. As Heron separated himself from the pileup during the play, a moose dressed in a Northbridge uniform bumped past him jeering, "I'm going to knock you clear to China on the next play turkey! Happy Thanksgiving"

With nine seconds left in the game, St. Mary's center snapped the ball to Heron who carried it around right end, twisted away from a would-be tackler, bounced off two more, and shook loose from a fourth. Moose loomed the last man between him and the goal, as Heron reached out with a stiff-arm, Moose's giant paws clamped his arm and a tug of war left Heron stopped just short of the goal line.

Just when all seemed lost and before the referee could blow his whistle, two Northbridge players slammed into the knotted players and all four toppled forward, the momentum carrying Heron and the ball into the end zone. Just as time ran out, Heron pulled the sword from the stone and fulfilled his promise to Azelia.

Just for you!

ON THURSDAY afternoon, bright sunshine highlighted three squirrels tracing jagged paths across the Community House lawn to escape to the safety of giant elms nearby, as Heron came walking up the street towards them. Off to his right, a charm of goldfinches in search of thistle descended onto the swaying branches of a mature lilac almost as one, a sure sign of the arrival of spring.

The red brick Community House with its white trim and tall white flag pole rising on its front lawn bore all the splendor of a country courthouse. Constructed in 1922 to shelter families of newly hired workers awaiting the construction of their homes, it evolved over time into a gathering place for Hopedale residents, complete with meeting rooms, theater, lounges, bowling alleys, and a gymnasium. Every Saturday morning children could attend feature films shown in the main hall at no charge.

No cars passed Heron as he crossed Dutcher Street and bounded the front steps of his family's modest two-story wood-framed house on the corner of Dutcher and Hope Streets, where the scent of his mother's cooking enveloped him even before he reached the front door. Inside, the modestly furnished house seemed comfortable in its simplicity. The front hall led to a family room on the right and a living room on the left, as well as a dining room, one bedroom and a bath. Stairs opposite the front door led to the second floor with two more bedrooms and a full bath. Another set of stairs in the kitchen led down to a full basement.

When the door closed behind him Emma wasted no time calling out from the aromatic kitchen in the rear of the house, her voice underscored with a sense of urgency. "The letter from the university arrived today, the one you've been waiting for. It's on the hall table."

"Notre Dame?" He felt his pulse quicken and his blood swell as he tore open the envelope and scanned the letter.

"Well?" she asked impatiently, when finally he entered the kitchen. "What do they have to say?"

He held it out to her. "Take a look."

Emma wiped her hands on her apron, smoothed out the letter on the kitchen table then read: *Congratulations. You have been selected to receive a full scholarship to the University of...*She swallowed hard. "Oh, my God!" The letter explained the school's academic requirements and expressed hope that Heron would soon join the ranks of the Fighting Irish.

"Oh Larry! I can't wait to tell your father! He'll be so proud, not that he isn't already."

He returned her hug and took a kiss on the cheek.

"What is it?" she asked, noting his look of dismay. "Why so solemn?"

When he merely shrugged, she added, "I thought this is what you wanted."

"It is, Ma. I...I'm just wondering how I could afford to pay room and board even with full tuition. Maybe they'll still want me next year. What smells so great?"

She returned to the stove and resumed stirring the simmering pasta sauce with that unnerved expression signaling she would not allow a change of subject.

Adding chopped garlic and a mixture of oregano, rosemary, parsley, and basil from two small bowls she prepared in advance, she selected a third bowl to put aside a bit more garlic to add near the end, so the flavors would peak when they sat down to eat. A fourth bowl held grated Parmesan for the table. "You must take these things when offered," she commented quietly. "They may never come your way again."

"We'll see." He scooped up a morsel of sauce with a wooden spoon, blew lightly then sampled a taste, the flavors exploding in his mouth. "Ummm! Mom! You've got to write down these recipes." She prepared food with a pinch of this and a portion of that, seldom with the benefit of a measuring cup or spoon. Without her cooking, the quality of life would sadly diminish.

Ignoring his request, she went on, "Don't worry, we'll come up with enough money, even if I have to take in wash."

"You would, too." He laughed and his powerful arms encircled her gently. "I really wish we could work it out somehow." With his parents' support, the scholarship, and the savings from his new job at the Draper Corporation, it just might be possible.

Suddenly the phone jangled alarmingly. "Heron residence," Emma sang cheerily into the mouthpiece. The usual smile that lit up her face whenever she answered the phone instantly gave way to a frown. "What? What? God no! Oh no!" Her face turned ashen and she gripped the two pieces of the phone so tightly that her knuckles matched the ivory on their upright piano. "How did it happen?"

After a distended pause, she said, "We're on our way."

"What is it mom?"

She collapsed onto a chair next to the kitchen table, her chin falling to her chest, while her body seemed to shrink physically into a vulnerable, defenseless ball, yet she remained composed and in control of her emotions. "Call a cab," she said, her eyes glazing over.

"What happened?"

With glistening eyes turned toward her son, she said in a whisper, "Your father. It was his heart. They rushed him to the Milford Hospital. Tried to save him but..." Her voice trailed. She grew more distant as her mind drew images of John waving to her from fields of ambrosia.

Heron bent to embrace his mother and felt her body quiver while remembering his father quoting Yeats: "In days of great joy, take comfort in the fact that disaster is just around the corner."

SINCE 1886, the Draper Corporation continuously maintained the Hopedale Cemetery, keeping its manicured lawns as neat and green as found on any plush fairway. Rhododendrons, lilacs, and hydrangeas lined pathways meandering past rows of untarnished grave stones, and majestic elms and clusters of lilacs provided just the right amount of shade and color.

Turnout for his father's funeral exceeded Heron's expectations; chief among those present, Fr. Edward Connors, whose words of condolence raised spirits and provided a temporary lift from the depths of despair.

After everyone departed, Heron remained by his father's casket silently asking forgiveness for breaking a promise he made one fall night as they returned home from the State Theater after watching "All American," starring Pat O'Brien as Knute Rockne and Ronald Reagan as George

Gipp. The highlight was Rockne's pep talk to his losing football team at halftime, years after Gipp had died of pneumonia.

"None of you ever knew George Gipp. It was long before your time, but you know what a tradition he is at Notre Dame. And the last thing he said to me, 'Rock,' he said, 'sometime when the team is up against it and the breaks are beating the boys, tell them to go out there with all they got and win just one for the Gipper. I don't know where I'll be then, Rock,' he said, 'but I'll know about it, and I'll be happy.'"

As they proceeded home, awash in the emotional aftermath of the movie, John said, "I want you to know how proud you make me, son. You play like a professional each week and just keep getting better."

"Love the game, Dad."

"I ran into Coach Morris at the barbershop last week. He told me that he has to drive the other kids but not you because you always push yourself to the limit."

Glowing with pride, his father added, "Someday I'll watch you play for Notre Dame. I can feel it. Hey, look at that!" He pointed at a falling star streaking across the heavens to disappear in the blink of an eye. "The Drapers brought us here because your brothers were exceptional athletes, and I expect that you'll carry on the tradition playing for the Irish."

"Maybe they won't want me - I'm only one quarter Irish."

Both men laughed.

Then John's voice took on a serious note. "Not to pressure you, son." He raised a hand as if to bat down any rebuttal. "I'd be proud of you no matter what team you play for, but after watching that movie I hope it's Notre Dame. It wasn't just the movie, I've always hoped to one day see you playing for the Irish."

"Don't worry, Dad. I'd never pass up an opportunity like that."

"That sounds like a promise?"

"That's a promise."

When he sat alone on his bed that night, Heron brushed tears from his eyes. Sorry Dad. But I know you'll forgive me because now I have to look after Mom. Last night he lay in this bed planning a future, but now survival held center stage. The money put away for college wouldn't carry them long. To keep their house and the low rent, he saw no choice but to pass on his dream and continue working for the Draper Corporation.

He lifted a newspaper clipping off his night table with a photograph of the Worcester County Sportsman soccer team snapped on October 3, 1926, just before the team boarded a ship that set sail out of East Boston for England. Draper players made up eight of the ten men lined up in their white soccer uniforms including his brother, John. Fred would have joined the men had he been of age.

His dad took such pride in that picture and would have been prouder still to see Heron play for Notre Dame. The future held no pictures of a proud dad standing next to his son in a Notre Dame uniform. Tears welled in his eyes as the full impact of the loss hit him.

John Heron, gone forever.

CHAPTER THREE

WINDS OF WAR

WORLD WAR II dominated the headlines as well as the hearts and minds of all Americans in 1942. In that year nationwide gas rationing went into effect and earlier in the month the U.S. Navy won the strategic battle of Midway. War movies ran one after another while increasing in popularity.

On Monday, Fr. Tom O'Malley entered Fr. Connors' office in the rear of St. Bernard's Rectory where he found the priest seated at his desk, deep in thought. On the wall behind him hung a framed diploma from Holy Cross dated 1927, and a certificate from St. Mary's Seminary in Baltimore where he earned his priesthood and ordainment.

"What's wrong Father?" O'Malley asked.

Connors cocked an eye and frowned. Usually, he welcomed his friend with an upbeat smile, a pat on the back and an invitation to sit down to a cup of his famous coffee.

"My God. We're losing our kids," Connors told him. "Our future."

White Cliffs of Dover drifted softly from a radio in a corner of the office where Connors served as associate pastor. O'Malley, who knew Connors as well as any man, crossed the room and laid the morning mail on the desk then sat across from him in an old but solid wooden armchair. "I know what you are feeling. So many dying, missing, or returning home wounded. These are terrible times."

The pile of mail suddenly spilled over, fanning out across the desk, the second letter from the top catching Connor's eye. Reaching forward with both hands, he tore it open expectantly, and as he began reading a smile tugged at his lips and crow's feet appeared at the corners of his eyes.

"What is it, Father?" O'Malley leaned forward, eager to share the news.

"Here. You read it," Connors said, grinning widely.

The letter came from the Most Reverend Thomas M. O'Leary, Bishop of Springfield. It read: *Dear Reverend Father: Your wish to tender your services to the United States as a War Chaplain has our approval. We have made known to the Military Ordinariate that you are seeking a Chaplain's commission with our knowledge and consent.* The letter further announced that Fr. Connors could leave his present job, declaring him eligible to take up his new duties as chaplain. He must write his intentions to His Excellency, The Most Reverend Military Delegate, Bishop John F. O'Hara then wait for instructions.

"How about that cup of coffee?" he asked Fr. O'Malley.

On July 15, 1942, Connors received his chaplain appointment from the War Department and proceeded to 207th General Hospital at Camp Livingston, Louisiana to report to the commanding officer on July 29. Soon thereafter, he wrote to his bishop as follows: *Thank you for the opportunity to represent the Diocese and to work with our young men. There are many wonderful opportunities in this life. I hope that I don't miss too many.*

On October 27, he forwarded a change to his APOE address, adding: *It is such a joy to go on various tactical formations with the men. It gives me a chance to become acquainted, and men are brought back to the church. Each day I am grateful for the priesthood. I have asked to go where the fighting is – and hope that I may be there soon.*

THEY NEEDED to get away, have some fun - forget about death, anger, and sadness. Azelia felt indeed proud of her job arranging the trip to Massachusetts' premier amusement park with its famous Nantasket carousel spinning under a scalloped canopy. Two rare Roman chariots carved by the Dentzel Company, each pulled by its own two horses, were voted the most beautiful carousels ever constructed.

As they walked arm-in-arm through the park, Azelia suddenly stopped to listen to words spewing from the mouth of a barker who pointed a walking cane at them. "I'll guess your weight within two pounds or

let you pick from this wide assortment of dolls and stuffed animals. Ten
cents. How 'bout you Ma'am," he called out to Azelia. "Step right up."

She laughed and turned to Heron. "Should I?"

"Why not?" Heron said. "'Course you know these things are
rigged."

"He won't come close to guessing my weight, you watch," she
whispered then turned to the barker. "All right, let's see you do it."

"Right over here, Ma'am." He surveyed her head to toe, so intensely that her cheeks flushed then proceeded to take her hands and pump
them up and down a few times while keeping his eyes tightly closed.
Heron flashed a smile. This was all part of the act.

The barker hesitated for effect then said, "One-hundred and six
pounds. Now if you'll just come over here."

Azelia leaned over and whispered to Larry, "Wrong. I weighed
myself this morning and I am one-twelve."

"Step right on these scales, Ma'am."

The monster scale wore a round face with a bouncing pointer that
settled at one hundred-seven.

"Ah, look at that! I guessed within one pound. Who's next? Let me
guess your weight. I will come within two pounds or..."

Azelia gave Heron an enigmatic look. "He cheated."

"What did I tell you?" he asked.

They rode the Dodgems, electric-driven bumper cars that smashed
into other cars loaded with strangers screeching in glee then eagerly left
their car and just as eagerly rode a 'chariot" through the Tunnel of Love.
When they exited, Azelia's face appeared flushed and Heron grinned
sheepishly.

"Test your strength!" shouted another barker who strongly resembled W.C. Fields. "How about you, sir?" he said to a man the size of
Paul Bunyan. "You look like you could win one of my giant pandas, great
teddies, or stuffed giraffes. Just ring the gong at the top." The pole was
marked WEAKLING near the bottom progressing to SUPERHUMAN at
the gong.

The big man paid a dime, rolled up his sleeves and raised the mallet high overhead. With all the strength he could muster, he smashed it
down and the red ball shot upward, reaching STRONG MAN, one notch
below SUPERHUMAN.

"Try again? Come on. You can do it."

The giant shrugged and disappeared into the crowd.

"How about you, sir?" the barker gestured at Heron, who shook his head to signify no.

"Why don't you try?" Azelia asked. "You want to, I can tell."

"It's fixed. That big guy couldn't do it."

"It's going to bother you if you don't at least try. I don't care if you don't hit the gong."

Heron handed over a dime and took the mallet in his hands with one eye surveying the barker to insure he did not signal someone. Then he wound up and brought the mallet down as hard as he could, shooting the red ball straight up the pole like a rocket to smash the bell so hard it left a dent.

Azelia squealed with delight.

The barker's face took on a sullen look. "Hey Buddy, you don't have to break the damned thing." Reluctantly, he handed Azelia the large stuffed panda bear that she pointed out.

The smile instantly returned to the barker's face as he turned toward new faces in the crowd gathered to watch. "See that folks? You too can win. Step right up and take your chances. How about you sir?"

Heron bought cotton candy and they crossed the street to walk along the beach adjacent to the amusement park, observing seagulls soar above breaking waves in search of food. Offshore, sailboats moved up and down the coast and a large freighter drew its outline on the far horizon. Heron came here in years past with his father to dig for clams, so very early in the morning they usually headed home long before the amusement park opened its gates. It seemed different now with throngs of people leaving footprints in the sand, walking, talking and laughing with seemingly no cares in the world. The sun never reached this high in the sky or shown so brightly on those early mornings with his father, when a heavy fog usually swept in to comb the beach.

A breeze suddenly caught her hair and whipped a strand across her forehead that she absentmindedly brushed back as she tilted her head against the wind. He could look for all eternity at her face, highlighted now by the sun, and never tire of anything so gorgeous and thoroughly captivating.

Her head lay on his shoulder and the panda on her lap when the bus pulled away from the beach. She couldn't be prouder. She might search the world and never find another man like him. To have found him in Hopedale in the early years of her life seemed improbable. For God to have made him so very special, so perfect, would always astound her.

Larry and Azelia skating on Hopedale Pond
Courtesy of Larry Heron, Jr.

ON MONDAY, July 6, the phone rang with a demanding intensity reminiscent of the day John Heron passed away. This time Emma remained solemn as she listened to the caller. "It's for you," she said to her son as he sat in the living room reading the Sunday paper. She returned to the ironing board beside the hot stove, moistened her finger and tapped the flat part of the iron to test the level of heat.

Heron heard the sizzle as he took up the earpiece. "Hello," he said into the stationary mouthpiece with the receiver pressed to his ear. The voice on the other end identified himself as a scout for the Boston Braves. "Mr. Heron, we've been monitoring your play for most of the season and invite you to try out for the team." Heron's sensational season with the Draper Corporation's semiprofessional baseball team and subsequent headlines regarding how he led the Blackstone Valley League in home runs and stolen bases dominated sports headlines throughout the region.

"We'd like to bring you to Boston as soon as possible. I think it's safe to say that in your case, the tryout is a mere formality. I am convinced we'll soon have you donning a Braves uniform."

Ordinarily, Heron would jump for joy, but this time he barely listened. No matter that the Braves could pay him handsomely, more in one year than he could make in five years working for the Drapers, no matter that this came as the answer to his prayers. The offer arrived one day too late.

"Thank you. Although I am honored, I'm afraid it's impossible. Yesterday I received my draft notice and report to the Boston Army Base on Wednesday, July 14th for induction.

Undaunted, the representative suggested Heron apply for a deferment. "No," he answered. "It wouldn't feel right, like shirking my duty. But I do hope the offer extends to the day I return."

The representative reminded him that Ted Williams received a deferment as sole support of his mom. Why not Heron? Heron followed the story closely in the newspapers. In the final analysis, Williams joined naval aviation and reported for active duty in November 1942. Heron left the scout with thanks but no thanks. First he had a duty to perform. The very next day the phone rang again and this time the caller identified himself as a scout for the New York Yankees. Heron gave the same answer.

Not only sports took a back burner; the marriage of Lawrence Heron to Azelia Noferi, planned for the following September must wait at least until his first furlough.

ON THEIR LAST Saturday together, the Ford automobile Heron borrowed from brother Fred followed the road as though it knew its own way. North on Dutcher, left on West Street then over dirt roads deep into the Hopedale Parklands past Maroni's Grove to the walk-in stone shelter built under a concrete roof with three stone walls including one featuring a huge stone fireplace at its center.

He parked a hundred yards from the grove near a clump of tall pines and they strolled along the causeway to the old Rustic Bridge, a setting Henry David Thoreau would die to preserve. The narrow Mill River rushed past boulders projecting from the water to form small eddies that swept under the bridge beneath them only to a rush to join a lake-sized part of the pond located on the opposite side.

They sat close together on one of two low stone rails on either side of the bridge, listening to the wind whistling through the pines and observed a variety of birds silently gliding through the evening haze en route to their nests. The shimmering surface of the water reflected the colors of fall leaves still clinging to masses of trees along its shores. Fish leapt here and there but each time they turned too late to catch more than a telltale ripple. Overhead a graceful hawk soared on a thermal in ever-narrowing circles without flapping its wings. Intoxicated by the strong scent of ozone in the air, Heron turned from the beautiful scene to place an arm around her. "Remember me when the stars come out to play and moonbeams skirt the bay. Remember me," he said.

"Where did that come from?"

"Not sure. Perhaps a movie, you know, the one where the departing hero tells his girlfriend to look up at the moon at night, and he will do the same. If both gaze at the same object, even from distant places no matter where on the face of the Earth it will somehow keep them in touch."

"But night falls at different times on the other side of the globe."

"Yet we will have looked at the same object on any given day."

"You're right, and when I look up at the moon I will tell it how much I love you." A shiver made its way through her, though the temperature stood in the mid-seventies.

Returning in a half-hour to Maroni's Grove, they stood beside the car listening to the sound of a rushing brook winding its way from a pristine spring further upstream and its travels past rocks and along a white sandy bed that ran past them. A gust of wind stirred then lifted a page from a discarded newspaper laying on a nearby picnic table. Suddenly airborne, the paper hovered like a manta ray before plunging to the ground to lay open at their feet. The headline read: *ARMED FORCES ALLSTARS BEAT BRAVES 9-8. Ted Williams' last inning homerun wins it for the All-Stars. Included on the list of All-Star players were Joe DiMaggio and Babe Ruth.*

A second gust flipped to a page that read: *GERMAN PANZER DI-VISIONS DRIVEN OFF; Allies unite bridgeheads and secure objectives.* The susurrus of war hung in the air and they could almost hear a clock ticking. The wind seemed to punctuate the words with an eerie whistle that sent a chill passing through them in turn followed by a rush portending something evil.

Heron tightened his arm around Azelia and took one last look around, thinking how much he loved Hopedale, its symmetry, peace, and quiet, and that this was not an easy place to leave behind.

ON THE MORNING of August 4th, Fred drove Heron, Azelia, and Emma to Union Station in Worcester under an emotional cloud made drearier by a sudden cloudburst. The procession entered the station wordlessly and followed Heron as he made his way toward the platform. The slightest whispers and the tiniest footsteps echoed off walls that stretched heavenward to meet the station's vaulted ceilings. When the Herons arrived at the platform, they spotted John Volpicelli, David Rubenstein with his mother, and Arthur Fertitta, occupying a long bench. They barely found time for introductions before the "all-aboard" sounded.

The train's engine built up steam preparing to leave the station without Heron, who remained on the platform holding Azelia in a tight embrace for one last kiss goodbye. "Hurry, or you'll miss the train," Rubenstein's mother shouted.

The wheels began clacking rapidly when Heron finally tore away from her in a sprint as it built up speed, It might have left him behind had David Rubenstein not reached out at the last second to take a hand and help him safely aboard.

Azelia wept unabashedly.

Mrs. Rubenstein placed a hand on her shoulder. "Don't worry, Azelia, he'll be fine. He'll come home to you." There was something comforting in the manner of Mrs. Rubenstein's words, while at the same time foreboding.

THE SUICIDE of a key employee following embezzlement at the post office created an ominous scandal and left an immediate opening. "We need someone bright, well organized, and capable of expeditiously tidying up the botched books," postmaster William Larson told high school Principal Winburn Dennett. "I thought you might know someone you could recommend."

Ensconced as high school principal seemingly forever in the town, Dennett knew every past graduate's capabilities intimately, and without hesitation answered, "I know just the person, a sharp, extremely ethical and trustworthy young woman who plans to marry soon, and could certainly use the money." Dennett promised to call her right away and explain that the job would last for only two months.

Azelia accepted the offer and kept the job for the next five months. Postmaster Larson wanted her to stay longer. "I only wish I could find more employees like you. You're extremely well organized and efficient. But I can't compete with Draper."

She thanked him for the opportunity but her job at the post office was done. Anyone could step in and maintain what she started. The Draper Corporation broke tradition by allowing women to backfill and make up for manpower shortages brought about by the war, and the company wanted the woman who recently sorted out the post office fiasco to join its payroll.

Larry and Azelia outside Worcester's Union Station.
Courtesy of Larry Heron, Jr.

Azelia happily settled into an administrative position in the tool design department where the routine kept her busy and took her mind off concerns about Larry's wellbeing for brief periods. Besides field hockey, she not only played basketball in high school but also proved an excellent softball pitcher, having captained the team her senior year. So it came as no surprise when she became a star pitcher for the Jigs, a Draper softball team that competed against the Main Office, Mechanics, and Shuttle, all-women softball teams with names related to looms and their manufacture. Had she been born a generation later, she might have played professional sports or been given a shot at becoming an executive in the company, but the women of her day found themselves fortunate to find any job openings.

CHAPTER FOUR

FT. RUCKER

ON AUGUST 9, 1943, the sun's rays slashed through clumps of stratus clouds to bleed a fiery red on the far horizon, at the same time shedding warmth on twin buses passing through the gates of Camp Rucker with fifty-two draftees onboard from New England, Wisconsin, Indiana and Mid-Atlantic states.

The twelve hundred grueling miles from New York to Rucker in the confined space of buses with no air conditioning and heat pouring down like air from a blast furnace made for sweat-soaked clothes and unbelievably foul odors. The record-breaking heat and humidity forecasted from Washington, DC to the Florida Keys the entire week promised no immediate relief.

When Pvt. Heron exited the first bus, he glanced down at the Bulova Azelia presented him on his twenty-third birthday to read 5:45 a.m., which marked fifty-five hours without a bath.

"Damn the southern heat," sighed William E. Shanahan from, Taunton, Massachusetts. "Damn the southern humidity, and the same for those stupid Barbasol, Nehi, and Dr. Pepper billboards. Didn't the south ever hear of Coke or Pepsi?"

Angelo Bastoni muttered as he sucked in a breath of fresh air, "Christ, if we could bottle your odor we could market it as a weapon and win the war." Then he bent to dig through a pile of luggage the driver finished stacking on the curb, looking for a small black duffle bag in which he packed his deodorant, adding, "Or market your sweat as an insect repellant."

"Yeah. We should name it Angelo's Revenge." Shanahan commented as he located the bag and withdrew it from the pile.

Raising his arms above his head, Steve Fiske, the farm boy from Chesterfield, Massachusetts said, "I'd be the first to surrender. Phew!"

"Phew yourself, lower your arms." Roger Burt quipped. "Or somebody please fart and clear the air."

"Ha. Ha. Real funny Burt." Fiske gave his arm a friendly poke.

Spoken words and laughter belied an undercurrent of nervous tension borne by fear of the unknown. These men had been inducted into an army about which they knew little, one that afforded them no pleasantries or the slightest control of destinies. Their lives suddenly interrupted, they tossed aside newly acquired jobs or plans to attend college to serve in the Army. Separated from the safety and security of hometown friends and relatives, these inductees found themselves on a path that could lead them directly to death's door.

The newly formed 87th Chemical Weapons Battalion, its principal weapon the 4.2-inch mortar, required plenty of acreage for teaching the art of lobbing shells at remote targets, plenty of open space. They found it here, in a place named Ft. Rucker, Alabama. By the end of 1941, the Federal government acquired possession of 65,000 acres of Alabama farmland, and last year the Corps of Engineers built a 4,600 acre cantonment naming it for Edmund Winchester Rucker, a confederate soldier from Tennessee. The stiff grass overrunning wide portions of Alabama, Florida and Georgia, easily mistaken for strands of copper wire became labeled "Wiregrass Country."

"I'm starving," groaned John Sears, from Plymouth, Massachusetts. Sears stood over six feet and moved with a lumbering gait. He resembled James Stewart, the actor who enlisted as a private in 1941 then rose to colonel within four years and would retire from the reserves as a major general in 1968. The ages of Heron's fellow draftees ranged from eighteen to twenty-four, so forty year-old Sears accurately told Heron on the ride to Alabama, "The army scraped the bottom of the barrel when they drafted an old fart like me."

"I'm ready for some grub," whined Burt, the no-nonsense twenty year-old ex-auto mechanic from Plainfield, Massachusetts. Forthright and likable Burt, who Heron first met at Ft. Devens, stood a wiry five foot seven with hazel eyes and brown hair.

Shanahan cocked an eye at Burt. "Yeah. Me too. Only I don't know which I want first, food or bath."

"You'd better eat while the mess hall's still serving." Heads turned toward the soldier who suddenly appeared from nowhere and whose coarse voice sounded like Sherman tank treads swiveling over loose gravel. "Welcome to Rucker, gentlemen. My name is Sgt. Volcjak. The sooner you finish role call and stash your gear, the sooner you'll eat. The army moves on its stomach, in case you didn't know, so fall in!"

Sgt. Carl C. Volcjak stood at six feet, thickly built with a square cut face seamed with lines of stress and browned by extreme exposure to the outdoors. An Army cap rode at a jaunty angle atop his military crew cut and his neat starched clothes showed fold marks as though recently unpacked from his footlocker. Promoted to private first class three months ago, he made corporal a month later, and sergeant in the month that followed.

Hands on hips, shoulders thrust back, head erect, the sergeant read names alphabetically from a clipboard, his sanguine gray-black eyes conveying sternness and competence. Just as the final "here" reached the sergeant's ears, the inductees heard a cannon roar in the distance. "That was Big Bertha. Get used to her because that's your wake-up call every morning from here on except Sundays."

"Let me brief you so you'll know what to expect. Training includes mortar practice and squad tactics, simulated support of infantry house-to-house fighting, jungle combat, and assaults on fortified positions. You will undergo strenuous calisthenics four times a week. The obstacle and physical fitness courses at the beginning and end of training will show how much you've improved. Exhaustive road marches with full field equipment will prepare you for battle conditions. Any questions?"

Silence.

"Grab your gear and fall in," the sergeant ordered.

The men lined up as instructed, still with a sense of extreme dislocation. It took just moments to stow their gear in barracks #12, a building unfinished inside with the ceiling a maze of ductwork and electrical wiring. Not exactly the Hilton but #12 would do as a place to rest after a hard day's work.

As they marched the short distance to the mess hall, the camp came alive with soldiers streaming from barracks like colonies of ants from disturbed nests. Sergeant Volcjak called roll, his voice clearly rising above the steady cadence of marching feet, as column after column marched smartly past in response to orders barked at them by the sergeant's counterparts.

Lined up outside the mess hall the new recruits resembled a bread line in the midst of the Great Depression and once inside, the food, though nourishing and substantial, proved not exactly five-star. Following chow the men lined up again outside barracks #12 to meet their company commander, Captain John T. Stiefel, who positioned himself upwind of the ragamuffin band that by now smelled like an old locker room filled with soiled laundry and yellowing sneakers.

The captain stood square-jawed and tight-lipped, his forthright blue eyes smiling in the manner of Fr. Connors. Despite the heat his starched shirt like the sergeant's, showed sharp creases, and though everyone here stood soaked in sweat, the captain's garb appeared neat and dry. Heron wondered if the captain lacked sweat glands.

"Good morning men, and welcome to Camp Rucker. I know you're anxious to wash and climb into clean fatigues so I'll keep it brief..."

The screech of brakes drew attention to the jeep slamming to a stop amid a cloud of dust ten feet behind the captain. An Army major hopped out and came forward while the jeep's motor continued to run. The captain about-faced and saluted Major James T. Batte, commander of the 87th Chemical Weapons Battalion (Motorized), which had been activated on May 22, 1943.

Ramrod straight and wielding a swagger stick that seemed an extension of his hand, Batte addressed the company commander with eyes like hard blue coals. "Why aren't these soldiers cleaned up and in starched fatigues?" The major wore full dress, his chest gleaming with medals, shoes spit-shined, and menacingly tapped the side of his calf with the swagger stick.

"Sir, they arrived by motor coach at oh five-thirty with no time to..."

"What time is it now, soldier?" Batte looked past Captain Stiefel as though he did not exist and snapped the stick jauntily under his left arm.

"Oh nine-hundred, sir."

"You have forty-five minutes to get them cleaned up and in proper uniform in time for indoctrination."

"Yes sir." The captain saluted as the major turned on his heels to depart then with a snappy about face ordered, "Fall-in back here at..." He hesitated to check his watch, "Oh nine-thirty," adding, "In combat fatigues and helmets." At that time the door to the jeep slammed shut and the major's jeep whisked him away.

Sweat moistened the captain's forehead at last and his mouth formed a tight thin line. The Pennsylvanian who had risen through the ranks impressed Heron by keeping his cool in the face of adversity, a good man to follow into battle. Though his face showed placid, the heat in the captain's eyes could fire a massive boiler. "Clean up quickly and get into starched fatigues," he ordered. "Look sharp. Dismissed."

After setting a new record by showering and dressing faster than ever before in their lives, the men felt fresh and smelled clean but the perspiring never ceased as they quickly reassembled for the three block march to arrive before a set of green bleachers. There Sergeant Volcjak ordered them to stop, right-face then fall out and take seats. As they moved onto the bleachers, a staff car pulled up as if on cue, and the driver climbed out to smartly open the rear door for Major Batte. The major exited the car carrying a leather binder in one hand, and with the ever-present swagger stick in the other, marched briskly to the podium, arranged some papers then turned to address the men.

As he looked up from the podium a hush fell over the stands.

"Men," he said, reading from the pages set before him, "Welcome to the 87th Chemical Weapons Battalion." A smile curled back his thin lips revealing even rows of sparkling white teeth but something else caught their attention, the stern eyes and rigid body language conveying all the warmth of a bronze bust in the heart of the Yukon.

"Here you will find excellent teaching facilities and competent instructors, amusement and recreational facilities. Your religious preference will be respected and our chaplains are here to minister to your spiritual needs." He paused while a helmet clunked down a few steps before an embarrassed private Tom Fletcher swept down to scoop it up. When the chuckles died down, he continued. "You will find your officers just in their treatment and anxious to teach. Many started out as privates who by hard work and diligent study have achieved their present ranks."

Major Batte's smile evaporated and his eyes seemed to pierce theirs one-by-one as he scanned faces in his audience. "You can do the same, but not by goldbricking, playing sick, or being a sloppy, inefficient soldier. If you are a Dumb Joe now, you will be a Dumb Joe when the war ends. By paying strict attention, you can emerge an efficient soldier, more likely to return from battle."

The major cleared his throat and continued, "The Fighting 87th has traditions to build. It is a chemical warfare battalion armed with the 4.2-inch mortar that fires projectiles containing gas, smoke, or high explosive."

He paused to scan faces for a reaction to the word "gas," and saw the usual frowns and turning heads. This group appeared even younger than the last crop, yet all seemed eager.

"Your mission is to render support to the infantry." His voice intensified. "You must strike quickly and with fury, like tough first class field soldiers. We don't give a damn how our gases and powder are made, or by whom; our job is to shoot them at our enemies, and shoot to kill."

Larry Heron at Ft. Rucker, Alabama, *Courtesy of Larry Heron, Jr.*

Now his voice softened and his body lost its rigidity, as if to punctuate the end his lecture and coast to a finish. "May you enjoy good health, happiness and good luck while you are members of the Fighting 87th. I know you are anxious to show those damn Germans and Japs that here is one battalion that can out-fight the infantry and out-shoot the artillery, and who are just as good and tough as they come."

The major knew he instilled just the right sense of what the 87th stood for, and that his words would ignite a curiosity as to the types of chemicals the Army planned for them to employ. Though just about everyone in the world knew that poison gas had been outlawed by international convention, these opening comments always raised questions and before training comes to an end, the men will find their answers.

IN THE FOLLOWING weeks and months, Heron would learn that the 1874 Brussels Convention outlawed the use of poisons and all weapons that might cause unnecessary suffering, and that the Hague International Peace Conference twenty-five years later added projectiles filled with poison gases to the ban. But when the Germans launched a major chlorine gas attack at Ypres, Belgium during World War I, all bets were off and both sides began using chemical weapons through to the end of the war.

The rationale behind the formation of the 87th became clear at a lecture given by Captain Stiefel on a balmy Wednesday afternoon. "The Treaty of Versailles banned chemical weapons but not their development, production or possession. All countries retained the right to retaliate in kind should an enemy first attack them or their allies with such weapons. Unless they do, we are restricted to the use of high explosive and white phosphorous, or 'smoke,' as WP is sometimes called."

"How long after the enemy initiates a poison gas strike would we get access to such weapons?" Private John Sears asked.

"Immediately," Stiefel answered. He paused for a moment then added, "Because we will carry mustard and phosgene gases into battle just in case, and let's hope we're not forced to use them."

Pvt. Heron reported to a squad leader with six other men, four gun squads to a platoon, three platoons to a company for a total of sixty-four men operating nine mortars. Each platoon had its own fire direction center and operated separately to maximize geographic coverage. When moving forward, one platoon would fire while another repositioned.

The squad soon learned to set up and fire in minutes with unbelievable accuracy. In time, they could drop high explosive rounds directly on fake fortifications, tanks, and targets towed to simulate an enemy on the move. The men fired white phosphorous that not only created smoke to drive an imaginary enemy from hiding, but also doused them with particles that would eat through their flesh. They learned to distinguish between differing types of shells denoted by gray or olive colors and coded bands of green, yellow, and red to indicate the particular chemical fill.

The 4.2-inch mortar barrel rifle bore caused the shell to spin like a perfectly thrown football, thus eliminating the need for a fin, and allowing for far more deadly accuracy than its predecessor, the old British Stokes mortar. The assembled mortar weighed 330 pounds. The heaviest of its three main parts weighed in at 175 pounds. The men were amazed at the ease with which Heron would single-handedly heft the heaviest part on and off the back of a jeep whenever the platoon advanced its position.

CHAPTER FIVE

A DIFFERENT DRUMMER

ON TUESDAY, following a tense day of practice firing, Sgt. Volcjak approached Heron. "The Major wants to see you in his office right away."

"What's up Sergeant?"

"No idea."

What did the commanding officer want with a lowly private he couldn't pick from a crowd? His mind whirled as Heron attempted to reject the thought that perhaps something happened to Azelia, or his mother, pushing that from mind, he strived to remain calm.

Heron's footsteps echoed off unpainted walls and linoleum floors as he walked past a tight warren of small offices in the one-story rambling barracks-style building, stopping when he spotted Batte's nameplate tacked on the center of a mahogany-colored door. He knocked and Lieutenant Bonafin closed the manual he had been thumbing through, inviting Heron to enter Batte's neat, sparsely furnished office. "Sit down," The lieutenant gestured toward a wooden chair then rose to go in search of the Major.

The scent of cleaning chemicals, most notably ammonia irritated Heron's mucosa as he sat waiting with anticipation in an office as antiseptic as a medical facility. The silver frame on the table beside him held a five by seven photograph of an attractive woman Heron assumed was the Major's wife. The walls on either side of the desk held a half-dozen plaques with Batte's name and dates inscribed on attached brass plates,

each surrounded by framed certificates of achievement and pictures showing various high-ranking officers pinning medals on the major's chest.

Heron's attention then shifted to a certificate showing that Batte
graduated from VMI in 1940, which fit with rumors that the major
dropped out of West Point as a plebe back in 1935. In every hatless photo
of the major, he sported the signature dome sparsely coated with strands of
hair that earned him the nickname "Old Knobby," a title never used in his
presence. Close to thirty-one, Old Knobby looked not much older than the
mostly twenty-year olds under his command.

I'll Be Seeing You wafted in through the thin walls from a radio
playing somewhere down the hall, and from outside the building came
a cacophony of engines, shouts of cadence, and the stomping boots of
soldiers marching past the open window. Heron also heard the pop! pop!
pop! of distant gunfire and the drone of a plane used for reconnaissance as
it passed overhead precisely when the door swung open and Old Knobby
walked into the room followed by Lt. Bonafin. The sudden opening stirred
a mild breeze that set the fringes of beige curtains dancing on either side
of the window behind where Heron suddenly rose to his feet.

"At ease, soldier. Please sit down." The major spoke dispassionately and flashed Heron a disarming smile. "How are things going with
you?" he asked.

"Fine, sir."

"Good. Good." The major moved around his desk to lower his
frame into a swivel chair. When standing, the major appeared taller because of his slim stature but now the oversized chair seemed to cut him
down to size.

Lt. Bonafin drew a chair even with Heron's and sat while the major
seemed preoccupied with an imaginary dust particle he absentmindedly
brushed from the shiny surface of his desk, emitting no signs that he had
invited the recruit to deliver bad news. Opening a manila folder, he drew
out a sheet of paper.

"I see you were an outstanding ball player." He read something
written on the paper then nodded his head affirmatively. "It says here
that when you captained the football team your senior year, St. Mary's
swamped Cathedral High shutting out Angelo Bertelli, Notre Dame's star
quarterback. Not only did you shut down the famous 'Springfield Rifle,'
you didn't allow him one first down."

Heron refrained from adding, I didn't do it alone, because he sensed the major was on a roll.

"Hmmm. Says you set schoolboy records for the most home runs and stolen bases in the Blackstone League, and since joining the army, maxed the aptitude tests and recorded best times on the obstacle and physical fitness courses." He glanced over at Bonafin. "What have I left out Lieutenant?"

The lieutenant twisted in his chair to face Heron. "Just that you were swamped with scholarship offers," he said. "As well as offers to join the Boston Braves and New York Yankees, which you turned down along with a deferment."

Heron felt amazed that these officers did so much homework, never imagining the army would run such a thorough background check. What else did they know about him?

The major remained solemn. "I'll come right to the point – Corporal Heron."

A look of surprise crossed Heron's face.

"Yes, you've been promoted. Keep up the good work and you'll soon make sergeant. But that's not why we invited you." Batte leaned back and rested his hands on the arms of his chair.

"I like what I see in your progress reports."

Heron shifted self-consciously.

"Lt. Bonafin made recommendations to Captain Stiefel who thought it better that I address you directly." The major leaned forward to place his elbows on the desk and tented his hands. "I won't pursue this matter unless you agree to go along." Lowering his voice conspiratorially, he added, "I'd like to sponsor you as a candidate for West Point." He leaned back. "Of course there will be tests. You'll have to compete with others but you certainly have the right stuff. The Army needs leaders, and from what I see of your high school academic records, you could easily make the grade." He paused to allow his words to sink in.

Heron felt deeply moved. Azelia would want him to accept, if for no other reason than to keep him from harm's way for a period of time. Smiling politely he cleared his throat. "Uh. Sir. I am honored that you think so highly of me...that you would recommend me. But I'm afraid I cannot accept."

The major made no effort to mask his disappointment. "Though you have the makings of a warrior, you also seem to have a history of turning down opportunities." A rasp crept into the major's voice that had been absent until now that contained a hint of annoyance. "You've refused scholarship offers from Notre Dame, Texas A&M, Trinity, Boston College, Holy Cross, and many other fine institutions." A pause. "And now West Point."

Heron cleared his throat. "There are reasons, sir."

The major hesitated then smiled grudgingly. "I'm listening."

Heron squirmed in his chair, not comfortable talking about himself. The lieutenant shifted his body in silent sympathy. "I would like nothing more than to play ball for West Point or Notre Dame." He looked down at his folded hands. "But it is not an option."

"Go on."

He told Batte everything, the promise to his father, saving up for their coming marriage, his father's sudden death, working to support his mother, and how the Draper Corporation rents houses to employees for next to nothing, even maintaining them free of charge, as long as someone in the family works for the company. "The benefits remain in effect as long as I am on duty, and will continue uninterrupted once I return to the job."

"I see. It seems you are responding to a higher calling and that's rather noble of you, young man. I might have misjudged you had I not known the circumstances." His voice grew softer, calmer. "It's a good thing for the Army if you elect to continue serving once the war ends. There are always battlefield commissions and other ways to become an officer. Like the Draper Corporation, the Army takes care of its own." He smiled with a practiced cordiality that did not extend to his eyes.

"Thank you, sir. I'll do my best," Heron promised.

Heron lost not time working on his commitment to perform. On Wednesday, he came away from the rifle range an expert marksman, and the following Monday led Tom Fletcher through the first of a series of exhaustive nightly drills to help the flabby lobsterman from Bar Harbor firm up. No one could understand how a man who made a living hauling lobster traps and performing heavy-duty tasks on a lobster boat for a living could not muster the strength to perform a single sit-up.

Heron's comrades admired his spunk. Tough, blessed with good looks, intelligence, and possessing the physical attributes men covet, he nevertheless remained down to earth. He treated every man as his equal, including Fletcher, the man who wore an apology tattooed on his forehead and seemed to hate himself for taking up space on the planet.

After a month of regular workouts, Heron had Fletcher knocking off ten sit-ups with relative ease and dropping ten pounds of flab on the way to building self-confidence.

Larry Heron, Camp Rucker, circa 1943
Courtesy of Larry Heron, Jr.

As his time at Rucker ticked past, Heron's bonds grew stronger with the men as they struggled together to survive the army's vigorous training and regimentation. Not only did they live together, they went into town as a group, got drunk and got into scrapes together.

ON A FOGGY Saturday in early December, Heron and David Rubenstein thumbed a ride into the town of Dothan, about twenty miles southeast of Ft. Rucker, where next to nothing went on except Christmas lights, so they made for a place called the Red Barn, where Pabst and Schlitz signs graced the front windows.

Stepping inside, they crossed a ratty wooden floor untouched by a mop for decades, past a Wurlitzer in the center of the room to a mahogany bar marked with beer stains. Confederate flags adorned both upper corners of a large mirror behind the bar alongside a sign beside it that read: "No Colored Served Here." Nodding his head to draw Rubenstein's attention to the sign, he whispered, "Let's make it a quick one."

Before Rubenstein could respond, a burly man on Heron's right with a hard, raw-boned face badly in need of a shave asked, "Where you boys from?"

"Rucker," Heron answered, an inner voice warning him not to mention Massachusetts.

"Let me buy you a drink, soldier," the man offered. It was not a question.

"Oh that's not necessary," Heron answered without rancor. "Thanks anyway."

"You refusin' my hospitality, boy?" Burly asked, pugnaciously. A towering man in his late thirties or early forties, he wore a black fitted T-shirt with short sleeves rolled up to display rippling biceps.

In the interest of civility, Heron replied, "Hey. No problem. You can buy us a drink if you'd like."

"I offered you a beer. Didn't say nothing' about Jew boy." He said it with his eyes fixed on Rubenstein's gold ring with the Star of David.

Heron tensed and looked at the bartender who averted his eyes, as suddenly it grew so quiet across the bar you could have heard one of the fleas crawling on the bar scratch itself. "Come on Dave. Let's go," Heron said.

Emboldened by the presence of three smirking friends, Burly said, "I would feel insulted if you was to walk out of here after I offered you a drink, boy." He turned an amused smile toward his associates who nudged one another and laughed encouragingly.

"Thank you," Heron said. "But a truck load of buddies returning to camp will swing by any second to pick us up." He looked at his Bulova. "Come on Dave, we better head out to meet them." He started for the door with Rubenstein close behind, but moving close on their heels, Burly reached up and grabbed Rubenstein's lapel, snapping a brief moment of silence with the sound of tearing fabric. "You don't leave 'til I say so." Burly's manner of speech now sounded petulant and intolerant.

"Take your hands off him," Heron ordered.

The redneck grinned, displaying yellowing teeth, pleased that Heron fell for his trap. "Suppose you make me, soldier boy?"

The slight and bookish Rubenstein struggled vainly to break Burly's grip, as the redneck gathered the other lapel in his free hand, pushing the shocked soldier onto a barstool. "Sit!" he barked then picked up a bottle the bartender just opened to begin pouring it down Rubenstein's front. The angry soldier reared up ready to fight when Heron took a tight grip on Burly's arm.

"Enough," Heron said quietly, with an inscrutable gaze worthy of

Charlie Chan, the fictional movie detective popular at the time.

"Sure," Burly said with a shark-like grin. He raised his left hand defensively while smacking the base of the bottle against the lip of the bar with the other, breaking off its end. Rubenstein sidestepped as Burly turned to face Heron. "See that sign?" he said pointing. "Jews, niggers, and northerners are not welcomed here." With the speed and menace of a rattler, he stabbed straight at Heron's face, an inside move with the jagged edge that seemed impossible to block.

Yet with unflinching and preternatural speed, Heron's left arm shot out to redirect the arm with the lethal bottle to one side. Both Heron's hands then closed Burly's beefy wrist like a vise, twisting it behind the thug's back to shoulder level with a final pull that sent the broken bottle crashing to the floor.

Burly's distorted face turned a shade of purple incongruent with human skin, and a strange guttural sound escaped his larynx, a sound that brought his two companions down from their stools. With Heron pushing his arm high behind his back, Burly doubled over whimpering in pain, no longer fearsome, and all eyes appraised Heron with new-found respect.

One man emerged from the crowd like a professional wrestler, cutting a figure the rival of Burly's. As he inched closer, Heron thought this is what Hercules must have looked like, a thick body with bulging biceps and no neck. His flat head bore a wartime crew cut, and his forehead, flat and shiny, reminded Heron of a bar brawler that once prowled the dark alleys adjacent to the Milford Hotel.

Just then, the bartender lifted his arms bringing into sight a double-barreled shotgun, which stopped Flat Head dead in his tracks. "Not in my bar," his raw voice urged, as he swung the gun around with a motion of dismissal. Nodding his head at the picture of a young soldier mounted on the wall behind the bar, he added, "My son, Mickey. Somewhere in Sicily last I knew. Now get the hell out of my bar and don't come back."

Heron released Burly who raised his arms in the posture of a supplicant, while to his friends whispered, "Wait until they leave. We'll follow and get them outside."

THE REDNECKS spotted the Jew thumbing a ride through the thick fog rolling in. "The other one must be taking a leak," Flat Head whispered. Burly crossed the street with his compatriots close behind and Flat Head bringing up the rear. No one took note as Heron moved swiftly in behind Flat Head. "Looking for me?" he asked, truculently. Flat Head spun around to take a swing at the figure behind him, finding nothing but air. Heron wore the forbidding look of a panther as he lifted a foot sharply up into Flat Head's groin, doubling the big man down to a manageable size, whereupon Heron dropped an elbow on the back of his neck, tumbling him to his knees.

A moment later, Flat Head reared back to his feet and rushed forward with both hands outstretched to take Heron by the throat. But once again he missed the target, Heron stepping aside then kicking Flat Head's legs out from under him. Cursing aloud as he hit the pavement, Flat Head soon found his right hand twisted behind his back to the base of his neck. Heron leveraged the arm to force Flat Head to his feet then ran him straight forward so that his head smashed into a rail post with a thud. Before he could regain his senses, Heron lifted him by his collar and pants belt then drove him ahead once more to crack a shoulder against a corner post.

"Jesus! My shoulder's broken!" the big man cried out.

Heron let go and turned to see Burly disappear into the fog as he ran down the road, his friends darting after him like hounds after a fox.

The excitement did not end there. Ten minutes later, a two-door red Lincoln convertible with a black top pulled up as they walked along the road, the passenger window rolled down slowly and the driver leaned across her companion to ask, "Where you boys headed? Need a lift?" The heavy-set bleached-blonde with short choppy hair and abundant makeup spoke with a southern drawl, her words slipping over her tongue like wet molasses on a cool day.

"Camp Rucker," Heron answered.

"Going right past it," the driver said. "Get in."

Heron pegged both women at about twenty-something.

The passenger, who stepped out reeking of cheap perfume, pushed her seat forward and rose to stand about five feet. Pushing back her brown hair with red streaks that fell like loose shades across her gray eyes, she stood back while the men climbed into the back seat.

Once the driver set the car in motion, Rubenstein asked, "Where y'all from?"

"Montgomery," she answered. "My daddy's the sheriff." She spoke the words as though her daddy were King of England. "Ever hear of Sheriff Thompson? He's quite famous," she beamed.

"Can't say I have?" Heron replied, glancing over at Rubenstein who merely shrugged.

"Should we know him?" Rubenstein asked.

"It's been in all the papers." The driver giggled. "You never read about the lynchin's?" She said it as though asking if they ever attended a carnival.

"What lynchin's?" Rubenstein gulped.

"Two niggers they caught with a white girl about six months ago, wasn't it, Melinda? It made all the papers."

Melinda replied, "Right around Easter."

"Like the nigger a few years back who raped a white girl in his shack. Her daddy called my daddy and...well, they have to learn."

The two men fell silent. Not what sheriffs did back home, although they understood such things did happen in the South. How could they not know, since five thousand such incidents occurred dating back to 1882.

Melinda asked, "Where y'all from?"

"Ah, Massachusetts," Heron responded.

Rubenstein repeatedly swiveled his head around to peer behind them looking for headlights.

"So what do you girls do – for work that is?" Heron asked.

"I'm a cosmetologist," Melinda announced flatly.

Silence.

"I work for a funeral home. Make up dead people," she added.

Dead silence.

She swiveled her head around with each comment, as if to toss her words at them above the dim of the Lincoln's engine. "It's no different than working on live people. Actually it's easier. Never get no complaints. Ha. Ha." She suddenly grew somber. "Down right spooky at times though. Like when you press the flesh, it stays pressed. Leaves marks though. Weird."

"Yeah, weird," Rubenstein gulped, as the two men exchanged knowing glances.

"It takes skill," she went on. "I can make them look just like the photos people give me."

Rubenstein said, "That's real interesting." He actually felt kind of sorry for her, wondering what other work existed for a young woman in these parts other than waitressing.

The two men took turns looking behind them for a motorized posse but none materialized. Finally and with great relief, they climbed out of the car a half-hour later thanking the girls for the ride then made it quickly through Ft. Rucker's gates.

"I'll bet they lynch Jew boys down here, too," Rubenstein said, as the Lincoln drove off.

"Welcome to Alabama."

Once past the gates of Camp Rucker, it felt like a safety net closed behind them, but for both men the south would forever conjure images of sharecropper shacks, confederate flags, and bodies swinging from ropes in dark shadows.

All future trips took them to Panama City, which Heron found delightful, and though not at all like the shores of New England, the beach sand and Florida seaside held a sweet charm all its own.

CHAPTER SIX

THE WEDDING

AS WEEKS rolled into months, Heron never lost sight of his dreams and aspirations, expecting in a few years he would return to sports. Meanwhile, he would keep in top physical condition, running laps at the track and lifting weights at the base gym.

On Thanksgiving Day, he reversed the charges after spending nearly an hour waiting to cram his wide shoulders into a tiny outdoor phone booth. Even with the door closed, he could barely hear her voice above the din of passing vehicles and the chatter of men queued up outside the booth.

Christmas Day found him especially lonely and spending it away from her extremely trying. After a heart-tugging talk with Azelia, he clocked five miles on the post track then lifted weights and threw punches at a bag in the post gymnasium before returning to the barracks for a steaming hot shower. At seven p.m. he left the mess hall following a hearty Christmas dinner, arriving at the barracks around bedtime to reread her letters by moonlight.

That night, thousands of recollections popped free in his mind, as he replayed happy years of courtship. Dating in the thirties and early forties included playing tennis on clay courts in the town park, ice-skating on Hopedale Pond, eating hotdogs and fresh popcorn on a Wednesday evening while listening to band concerts in the park, or taking a stroll through the park hand-in-hand, with the scent of freshly–mowed lawn on a clear night while looking at the stars and holding her hand.

On Christmas Eve the barracks held a party around a tree placed in the middle of the room that looked like a reject from a paper factory. The men each contributed to the decorations, garnishing it with colored paper, garlands of popcorn, and silver balls fabricated from rolled cigarette foil. John Sears dressed like Santa, creating much laughter and cajoling as the men wished aloud for items like a red fire engine, a night with Betty Grable, an end to the war, a month's furlough, money and, of course, liquor.

Santa handed out gifts bought with donations, some useful, others not, like cigarettes, candy, prophylactics, beer, Vaseline, trusses, Kotex, and foot-long toothbrushes. Meanwhile, Burt, Healy, and Bastoni handed out beer that Cookie kept stashed in the cooler in cases marked Evaporated Milk.

The fun and games came infrequently and in small doses and by the end of January, the 87th became a crack outfit proficient in employment of the deadly mortar.

At 0600 on February 1, the battalion departed Rucker on buses that delivered them to Camp Forrest, Tennessee nineteen hours later for phase two training. Early the next morning, each man dragged his carbine under barbed wire with live machinegun fire passing inches overhead and charges exploding close by to simulate actual combat conditions. On the 12th, an Air-Ground Liaison Test administered by the 11th Detachment, Special Forces, tested the men with coordinated attacks that included live artillery and coordination with infantry units. At the completion of this exercise, the 87th received a citation for outstanding performance and the major pronounced them ready for action.

On February 13, Heron along with 449 officers and enlisted men departed Camp Forrest on a furlough cut back in the eleventh hour from two weeks to eight days. Something big was planned but something more immediate and important awaited Heron.

When she appeared in her wedding dress, his stomach turned somersaults. No amount of rehearsals could have prepared him for how much more beautiful she appeared now than he remembered. With hair pulled back under a simple yet elegant white wedding crown that accented her bright blue eyes, she looked to Heron like a movie star. Her sure-footed stride down the aisle and elegant figure accented by a fitted gown added an almost iridescent glow as she moved down the aisle towards him with a train following regally behind her.

As they stood side-by-side taking vows, she found herself trembling and full of doubt. Does he like me in this gown? Will I disappoint him on our honeymoon?

Hopedale Sacred Heart Church Wedding, September 5, 1942
Courtesy of Larry Heron, Jr.

The Milford Hotel, regarded by many as a less than desirable place for a reception, on this day looked beautiful with masterful decorations featuring ribbons of white and bouquets of flowers that added sparkle to each table. Toward late afternoon, the bride and groom glided across the floor to the strains of Stardust, like the only two people on earth.

Part way through the night, she shook out her hair and let it fall to her shoulders, smooth and clean smelling. Sparks flowed between them like electric currents when he held her close, the fragrance in her hair more intoxicating than all the drinks he consumed.

"The Milford Hotel worked out just fine," she whispered in his ear as they danced. "No one likely will ever forget our wedding."

When her mom first suggested this place of darkness for the reception Azelia stood aghast. How could she think of holding it in an old building that emitted the strong scent of stale beer, in a room with floors covered with a permanent film of dust and drinks served on a long mahogany bar marked with circular stains? But tonight somehow none of that mattered for the food tasted great with the price just right because Joe Sardini, the owner, offered it at no cost out of a long friendship with her father.

"Your hair smells of jasmine," Heron whispered in her ear. Though lacking an idea of what jasmine smelled like, he had read about it in books and heard it mentioned in movies. Tonight, it smelled heavenly and set his heart pumping faster, so it seemed the right thing to say.

"You smell good too," she remarked, leaning her head against his cheek.

He kissed her hair and thrilled to its softness.

And so they became man and wife at long last. Two people clinging to one another like swimmers lost at sea, knowing not what the future held in store.

THE MOST EXCITING wedding gift came from Heron's brother Fred who must have suffered weighty pangs of conscience for allowing his kid brother to discard a bright future so he could single-handedly support their mother. He gave the couple his old car and bought a new one for himself. Heron never asked or expected help from brothers Fred, John, and Leonard, who long since married and moved on but it pleased him just as much to take ownership of the old Ford as it did Fred to drive away in a brand new automobile.

Following Heron's instructions, Azelia sat at the wheel adjusting the choke and throttle then turned the on ignition while Larry bent in front of the car to crank the engine. It started on the third try and off they went to start their honeymoon in this car with its arthritic engine.

Heron found it difficult to take his eyes off the remarkably beautiful woman seated beside him, and drove along the highway the proudest and happiest man alive. To the pedestrian, Azelia seemed cheerful and bright but Heron knew that beneath the surface lay an undeniably strong woman with just a hint of vulnerability, a sexy combination if ever he knew one.

The couple spent their abbreviated honeymoon at a friend's cottage in Chatham on Cape Cod under sunny skies with balmy breezes. The first afternoon upon their arrival Azelia remained anomalously silent as she unpacked and put clothes into drawers, their beginning moments a bit strained and awkward. It seemed to take forever before she emerged from the bathroom that night arranged in a white silk robe that clung to her exquisite body like a second skin. Even without makeup, her features matched perfection, her nose perfect, lips full, intense blue eyes warm and exciting.

He stared into those eyes for another eternity before blurting, "Er... want to go through this list of restaurants Johnny gave us?" He seemed nervous and fidgety as a kitten in a den of wild dogs.

"Later," she said taunting him with half-closed eyes and a wicked smile that surprised even her. Deliberately, wantonly, she stripped the covers and stretched out on the bed. Her hair cascaded around her shoulders and her robe fell open revealing long lithe legs, the muscles flexing in a way that set his heart pulsating. Sports had served her well, body taunt and fluid yet strong and vibrant.

He took his time looking at her smooth chin, full and exotic lips slightly parted, teeth even and white then gently then kissed the hollow of her neck. "I love you, Larry Heron," she whispered. "Always have. Always will."

"I love you, too, Mrs. Heron," he said tenderly.

Such moments mimic dreams but dreams don't always repeat and seldom survive the test of time.

THAT NIGHT they drove to an Italian restaurant under a bright moon drawing in breaths of air with the fragrance of a brisk afternoon shower. They started with a glass of Champagne at the bar then moved to the dining area where he ordered a bottle of Chianti. When the waiter returned to take their order, she selected the insalata caprese and fettuccine Alfredo while he chose a small antipasto and chicken cacciatore.

Light from a candle set off her high cheekbones and delicate chin, accenting her glowing hair. When she caught his admiring glance, she leaned forward and gave his hand a loving squeeze. Come Back to Sorrento drifted from speakers in each corner of the room and people tossed knowing glances, as though they resembled illicit lovers instead of husband and wife. When they finished, the owner stopped to ask if they enjoyed the meal.

"We loved the chicken cacciatore," Heron said, pouring the last of the wine. The garlic-laced smells emanating from the kitchen lingered in the air so strong that he felt he could eat another helping.

"And I helped him finish it," Azelia added.

"It was my mother's recipe – my favorite. I'm so happy you enjoyed the meal." He bowed and disappeared into the kitchen. Moments later he emerged carrying a new serving of cacciatore and a pitcher of red wine, and placing both on the table said, "This one's on me." It added just the right touch and both left the restaurant later that evening in an exceptionally cheerful frame of mind.

She awoke the next morning and felt empty space next to her. Alarmed by the quiet of her surroundings, she called his name but after checking each room found them all empty. Tossing on some clothes, she carried her search outside and along the beach where the ocean lapped the shore then further along to where she spotted him jogging towards her. If this were summer and fifty degrees warmer, she surely would have found him swimming in the ocean, cutting through the water like Johnny Weissmuller star of the Tarzan movies, for he loved the water and would swim at every chance.

A hundred yards out, she spotted a large red buoy rocked by choppy waves sounding its bell, which drew her eyes to a half-dozen seagulls circling above a small fleet of fishing boats headed out to sea. She inhaled the freshness of the salty air and wondered why every day could not be as simple and serene as this one.

After he showered and shaved, he bundled up and stepped out on the front porch while she took her time getting dressed. Before she came out to join him, he moved to the porch swing, rocking gently as he watched gulls bobbing over the currents while making their familiar noises. Then he looked far to the west at puffs of charcoal cotton clouds in the sky and thought of war and death and man-made thunder fast approaching. And suddenly, an ominous dread fell heavily upon his shoulders and he could hear his own heart beating disquietly.

THE NEXT MORNING opened with bright sunshine and unseasonably warm weather for the time of year, so she slipped on a pair of slacks and threw a heavy sweater over a plaid blouse. He wore corduroys and a sweatshirt pulled down over a flannel shirt, just enough protection to ward off the chill as they strolled along the beach listening to the shrieks of gulls overhead and the waves tumbling lazily onto the shore, a serenity of surroundings under the warming rays of the sun, eerily at odds with the chaos of a world gone mad.

Heron snapped yet another picture of her sitting on the bow of an overturned rowboat, catching her with strands of loose hair outlining her face with glowing highlights beneath the sun's rays. Morning passed quickly, and as noon approached they settled on a bench in companionable silence taking in the sights of lobster boats and sailing vessels gliding across the bay while seagulls moving in unison fanned the air. As she turned to face him, he gently cupped her face in his hands and felt currents of electricity pass through his fingers to the tiny nerve centers of his body. The sweetness of her was like a narcotic to which he felt completely addicted. No one should be this happy, he thought.

Azelia (Noferi) Heron
Courtesy of Larry Heron, Jr.

They decided to eat in that night. Azelia cooked a small pot roast to help extend the week's meat rations. Parts of their wedding presents included a linen tablecloth, napkins, fancy silverware, and lit candles that she used to adorn the dining room table. They drank David Noferi's homemade wine and listened to Frank Sinatra records. After diner she danced with Heron holding her close and eyes closed while they moved as one, with the war on temporary hold, replaced by more immediate needs and emotional lapses to where the mind prefers to dwell. It lasted until the record player dropped a new platter onto the turntable and the words of the next song belied the mood of the evening, "Dream when you're feeling blue…"

Heron poured chardonnay into two glasses, setting both down on a coffee table in front of the sofa where they sat to gaze out though a large picture window at the seascape. Moments later, he leaned close to inhale the sweet musk of her like the intoxicating aromas of a newly opened bottle of wine. "The scent of your perfume fits you perfectly," he said. "Innocent, perhaps fecund, with just a hint of carnal desire."

She laughed mellifluously and kissed him on the lips. "Where did you learn those fancy words?"

"Never mind that now, Scarlet." With brows furrowed and lips pursed in a perfect imitation of Clark Gable, he said "Scarlet, I want you to know that I do give a damn! Let me prove it!" With that, he lifted her in his arms and carried her to the foot of the stairs leading up to a tiny bedroom.

"Put me down this instant Rhett Butler," she laughed, summoning a mock primness into her voice.

"In my own good time," he shot back while twitching his lips and shifting his jaw, à la Clark Gable.

Up the stairs he carried her and into the bedroom where he placed her down gently on the bed then climbed in beside her. They both laughed and then he kissed her tenderly.

"Oh Larry. I love you so much. I wish you didn't have to…"

"Shhh," he whispered and kissed her cheeks tasting salt from fresh tears. "I love you more than words can tell."

THEY ARRIVED in Hopedale early the next morning, the day ending with tearful goodbyes at Worcester's Union Station with Heron running once again during the final seconds to catch his departing train. Moments later, he sat gazing out a window as the train gathered speed. Leaning back against the headrest, a great sense of loneliness seized him, and an icy chill snaked down his spine. Scrunching his chin down into the heavy collar of his overcoat like a turtle withdrawing into its shell, he closed his eyes to allow a part of his mind to relive the tender moments they shared on their brief but wonderful honeymoon, while another section of his mind listened to the repetitive clacking sound of the train transporting him ever closer to his date with destiny.

On February 22, he arrived back at Camp Forrest for the start of further specialized training and the next day, newly promoted Lt. Colonel Batte resumed command of the battalion. By the 25th, the strength of the outfit reached 624 enlisted men and 41 officers, and their days overflowed with intensive training and repeated inspections.

A month later, just before the entire battalion departed Tennessee via rail for a permanent change of station to Camp Shanks, New York, he managed to sneak in a hasty call to Azelia to update her of his whereabouts. "I've been granted a weekend pass," he told her excitedly. "Can you meet me in New York City on Friday?" I checked ahead and there are plenty of rooms available at the Claridge."

"I can't wait!" she told him, her voice brimming with joy.

IN 1944, the tower that cast an incongruous shadow over Union Station in Washington Square represented the tallest structure in the city of Worcester, Massachusetts. Resembling Florence's Palazzo Vecchio, it stood as the last vestige of the old station, demolished in 1909. When completed in 1911 at a cost of $750,000, the new Union station took the spotlight as the grandest train station in the nation. Twin terra cotta domes rose on either side of the main entrance filled with rows of stained glass that filtered a myriad of colors over ornate sculptures ensconced below; and set between the massive domes stood a giant canopy of arched ceilings that sparkled like enormous jewels in the sunshine.

Thursday, March 29 marked a day Azelia anxiously awaited since the time she received his phone call. Arriving at the station, she found the main floor lined with rows of dark wooden benches amid throngs of people gathered there. Most were young men in uniforms who hugged and kissed family members and close friends, while most women shed tears - reminders of the day Larry departed from a platform on his return to Tennessee. So absorbed was she that day, she failed to notice the many popcorn vendors scattered about the station, or the shoeshine stands, or Mr. Podbielski's famous barbershop where five barbers stood busily cutting hair. Only this time Azelia awaited her turn to depart for New York City for a glorious few days with Larry before they shipped him to God knows where.

Once aboard the train, she folded her heavy overcoat and laid it along with her scarf and gloves on an empty seat to her right then sat down and bent to undo her boots. As she wiggled her frozen toes waiting for feeling to return to them, she glanced across at Angie Volpicelli sitting down beside Angie's mother-in-law, Alvina Volpicelli, the woman who earlier insisted on going along to insure her daughter-in-law's safety.

On the floor beside Alvina stood a large shopping bag overflowing with boodle for son, John. The heavily laden bag held salami, cheese, wine, and a variety of spicy foods that filled the entire train compartment with smells reminiscent of Boston's North End, where her mother used to take Azelia shopping for Italian foods and spices several times throughout the year.

The compartment felt cold, so Azelia lifted her coat from beneath those the Volpicelli women piled over it and wrapped it over her shoulders to keep warm. Her two companions followed suit then resettled into the two seats directly across from Azelia. Once the locomotive picked up steam and the train began to sway gently in rhythm with the clickety-clack of its wheels, most passengers, including Azelia's two companions, began to doze. Azelia too closed her eyes but could not fall sleep, not simply from the excitement or anticipation of seeing Larry, but more because of a nagging premonition that something would somehow go terribly wrong.

There seemed no reason for concern, for Larry remained safely settled at Camp Shanks in Rockland County, Orangeburg, a staging area for troops headed overseas from New York ports of embarkation. The Army promised him a three-day pass, a nice chunk of time during which they would enjoy one another's company one last time before separation, a meeting both had been looking forward to with great excitement.

To release her mind from further apprehensions, she began reading an article in Photoplay magazine about a famous movie actor who stood accused as a Nazi sympathizer, and thus occupied she ever so slowly succumbed to the motion of the train. Her head nodded then she fell in and out of a light sleep that lasted until the train lurched to a sudden stop at Grand Central Station in the heart of New York City, jarring her fully awake and ready to disembark.

The scent of spring filled the air as the three women stood outside hailing a cab that delivered them to the Claridge Hotel in the heart of New York's Times Square. They checked into a room that on the surface seemed neat and clean, though poorly lit. Against the wall between two beds stood a sink but the toilet and shower were located in a bathroom down the hall, something no one had anticipated and the hotel failed to advise prior to their arrival, which explained the low cost of lodging.

As evening approached, the room grew steadily darker and when

they left the room to make their way to the elevator, the halls were so
dimly lit they could barely find their way. When the women reached the
lobby, Azelia asked the concierge for names of restaurants he might care
to recommend. A block away they found the delicatessen he claimed carried good food and where they could eat a meal cheaper than at a standard
restaurant.

No one heard a word from the men that night.

The following morning they ate breakfast in the hotel and waited
four hours in the lobby, but neither Larry nor John arrived, and still no
messages. There seemed absolutely no way for the women to reach their
men. At noon, Mrs. Volpicelli insisted the girls leave to grab some lunch
and asked them to return with a sandwich while she held court in the
lobby.

"I'M SORRY, Cpl. Heron, but like I told you, all furloughs have been
canceled," Sgt. Julian Brunt from Mississippi, told him.

"But my wife is waiting for me at a downtown hotel right now
sergeant. I was guaranteed this pass."

"The Army guarantees nothing corporal, especially during a war.
Colonel's orders. Nothing I can do about it."

"But my wife!"

He shook his head. "Sorry."

"Let me at least call the hotel to let her know."

"Can't allow it."

"Why not?"

"No one can call out because the base is locked down and everyone in it secured."

"What about the guys still out on leave?"

"They've been ordered to return immediately. All they know is to
get back here pronto."

HE LABORED until he felt his lungs would burst and his joints flashed
fatigue alerts. Instead of stopping to take a rest, he turned up the heat,
increasing his speed, running as fast as his body would allow, pushing
harder than ever before in his life. He ran with fervor, as though punishing himself for not keeping his appointment to meet Azelia at the hotel
or sneaking off to locate a phone just to tell her how much he loved her.
But making a call might lead to confinement for the next six months. Sgt.

Brunt had warned, "If anyone breaches security, I will make it my life's work to see that he ends up in the stockades surrounded by barbed wire with a shotgun up his ass."

One side benefit, he thought amusingly, is that besides letting off steam, hard running helped rid his body of empty beer calories, which lately he consumed at an increasingly high rate. Sweat stung his eyes blurring his vision, as he pushed so hard he found himself sucking air through his nostrils then letting it out through his mouth in harsh, shrill tones. His heart and legs pumped in unison as he bested all prior running speeds and levels of endurance. But nothing, not even the punishment he gave his body could relieve the anxiety. I must find a way to reach her.

Showered and dressed, he headed for the recreation room to solicit advice from friends when Staff Sgt. David Thomas stopped him. "Lt. Bonafin wants to see you right away."

Something happened to Azelia. No, that can't be it. Word doesn't travel that fast. Perhaps his orders changed and they would grant a weekend pass after all.

He followed the sergeant to a small building, drab and poorly lit, where inside he found the lieutenant smiling up at him as he entered the temporary office. Bonafin offered Heron a cigar and told him to take a seat. He took the cigar, thanked the lieutenant and then slipped it into his shirt pocket. Pulling up a metal folding chair, he sat down and looked from the lieutenant to Sgt. Thomas, who wore a huge grin.

"Congratulations!" said the lieutenant. He hesitated then finished, "Sergeant Heron." The lieutenant's words provided a momentary lift, but Heron's heart remained heavy with concerns for Azelia's well-being. "Thank you, Lieutenant. Thank you Sergeant." He feigned a smile but felt nothing inside.

Although a likeable enough fellow, Sergeant Thomas was still a New Yorker who poked fun at Roger Burt about the Yankees beating the Red Sox. Nevertheless, Heron admired him for his street smarts. Like John Sears, whatever the battalion needed, Thomas somehow found a way to come up with on demand. Back at Rucker, when Lt. Bonafin needed paint and brushes to spruce up the interior of the barracks, Thomas "requisitioned" ten gallons of paint and a dozen brushes. No one knew from where and no one asked.

After the lieutenant departed, Thomas invited Heron and Roger Burt to the NCO club and bought the first round of drinks. Heron smoked the cigar and entertained the notion of drowning himself in beer as fellow noncoms continuously stepped forward to offer congratulations then ordered a new round. Because he decided not to share his problems and instead kept them locked up inside, his fellow noncoms watched him down beers one after another and attributed the consumption to elation over his promotion.

As Sgt. Robert "Red" Meyers stepped up to shake his hand and order yet another round, he asked, "Have you ever been to sea?" The sergeant's eyes twitched and he fidgeted with his glass as he spoke.

"Lots of times. My father used to take me fishing in Quincy Harbor and York, Maine," Heron replied.

"Ever get seasick?"

"We got hit with some pretty frightening seas at times but it never seemed to bother my stomach."

The sergeant's fingers tapped nervously on the bar. "The mere mention of the ocean makes me sick. I'm not looking forward to the long boat ride," he groaned. Meyers had a faraway look in his eyes and his face wore as tortured a look Heron had ever seen.

"Mind over matter," Heron offered his red-headed friend. "It won't be so bad on a large ship like the Queen Elizabeth."

Meyers' face turned ashen in reaction to the word "ship," so Heron thought it best to change the subject.

"Where are you from?" he asked.

AT A FEW MINUTES past six p.m., the desk clerk approached the three women sitting in the lobby to ask politely if he could be of assistance.

"No thank you," Azelia smiled. "We're waiting for our husbands."

At seven p.m., Angie told the desk clerk they would return by nine in case anyone came looking for them. Then the women left to dine at the Waldorf Cafeteria.

Nine-thirty found them back in the hotel lobby and when they checked, found that no one came looking for them nor did they receive any messages. The women waited until 11:30 before retiring for the night, leaving a message with the desk clerk to notify them immediately should anyone call or come looking.

The next morning after breakfast with still no word, people passing by a third or fourth time began to stare at the three women who seemed lost with nowhere else to go.

"What should we do?" Angie asked.

"We'll wait." Azelia answered quickly. She hadn't come all this way to return home without seeing Larry.

At four that afternoon, a man wearing a dark blue suit and a gray fedora approached with an intensity about him that caught Azelia's immediate attention. She thought he could have stepped out of a G-man movie. Cop was written all over him as he removed his hat and took a seat in the wing-backed chair directly across the coffee table from them.

"Excuse me ladies," he said. "My name is Philip Rothman, and I'm the hotel detective." He cleared his throat. "I could not help but notice you waiting in this lobby the past several days. Do you mind if I ask why, or who you are waiting for?"

"Our husbands and her son," Azelia said, nodding at Mrs. Volpicelli. "All servicemen stationed at Camp Shanks."

"They were supposed to meet us here by yesterday morning but haven't shown up or called," Angie added anxiously, hoping the detective could help.

"What do you suppose happened to them," Mrs. Volpicelli asked. "Do you have any idea? Is that why you decided to talk to us? Has something happened to my son?"

"I'm afraid it means they're not coming," he answered truthfully. "I've seen it before. My advice to you ladies is to go home – wherever that may be." He sighed and leaned forward wearing a frown. "If they haven't called for the past few days then it means they are on a ship headed overseas."

"Wouldn't they leave word?" Mrs. Volpicelli asked.

"That's the way it works sometimes, Ma'am. They can't let anyone know when they are leaving and must keep the destination top secret. You could be here for the duration they will not show and you won't hear a thing from any of them. The Army brings 'em to Shanks for one purpose, get 'em aboard and ship 'em out. You had best head for home ladies."

Reluctantly, the three women boarded a train the next morning for Worcester, Azelia teary-eyed most of the trip. Larry was headed into harm's way without so much as a final embrace or a goodbye kiss from her. She wanted to tell him just one more time how much she loved him but could only pray now for his safe return.

HE WOKE WITH a start and rolled to one side in his cramped bunk to peer into the darkness wondering if he really came awake or did he have one foot still in the dream with her smooth, naked body curled against him? Her soft smile lingered like the afterglow of a flashbulb fired in a dark room. He closed his eyes with the hope of reentering the dream long since absorbed by the abyss called reality. It took a moment to realize that he lay aboard the Queen Elizabeth as it plowed through a roiling sea, transporting him closer to his appointment with destiny.

CHAPTER SEVEN

UTAH BEACH

ON THURSDAY afternoon, April 6, the Queen Elizabeth steamed up the Firth of Clyde to drop anchor off Greenock, Scotland, where cheering British subjects surrounded a band of Scottish pipers in kilts who stood by to welcome the Americans with the stirring drone of their instruments. At 0800 they soldiers boarded a train that delivered them through dense fog and rain to Tiverton station in Devon County, a popular peacetime holiday destination marked by rolling hills and villages populated with eye-catching thatched cob cottages.

For the next seven weeks the soldiers devoted daylight hours in the quaint riverside town to training exercises and battle preparations, but the schedule proved far from a case of all work and no play. Weekdays ended with nightly pub visits and on weekends the men accepted invitations to dine in British homes on fish and chips sprinkled with vinegar and rolled in newspapers. And just as quickly as training exercises began, they came to an end, and at 0830 on June 3 each man received Hershey bars, extra cartons of cigarettes, a French-English phrase book, special French francs, and real bullets to load into their carbines.

No more games played with empty weapons on friendly fields. Instead the men loaded their gear into the backs of two-and-a-half-ton trucks that bounced them roughly over British cobblestones toward Plymouth. From time to time, Heron would peek out the back at an England transformed almost overnight into a vast ordnance dump with mountains of stores stacked in previously barren fields. Fifteen minutes later the trucks turned off onto hastily constructed roads marred by rusty soil and corrugated tire chevrons in order to avoid the slow crawl behind rows of trucks, ambulances, tanks, armored cars, jeeps, bulldozers, and pedestrians on bicycles.

Roger Burt gripped his French phrase book tightly in both hands each time the truck swerved to negotiate one of England's classic sharp curves or steep inclines. To pass the time, he began practicing on his friend, Francis Healy, who sat opposite him in the rear of the truck. "Parlez-vous Français?" Burt shouted above the cacophony of grinding engine and metallic groans from beneath the truck bed that rattled his voice and made him sound out of breath.

"Voulez vous couche avec mois? That's all the French I need," Healy responded.

"Wants put to bed you with months? That'll get him far," Heron admonished.

"Ha, ha. Funny Heron."

"How does an Irishman come to know French?" Fiske asked.

"One-quarter Irish," Heron corrected. "I studied it in high school."

Burt laughed. "Is that what they teach you in Catholic schools, how to negotiate your way into bed?"

Heron pictured Burt and Healy as a matched pair of short wiry bookends. If God had made them six feet tall, they could have whipped the entire German army on their own.

Although joking and laughing during earlier verbal exchanges, the men now rode quietly, each heavily burdened by an ever-present and steadily increasing undercurrent of fear. After seven intensive weeks of preparation in Tiverton, four hundred-forty officers and men comprising the 87th assault wave finally made it into sixty-seven vehicles transporting them minute-by-minute closer to harm's way.

Removed from clogged traffic, the trucks attained a constant speed, causing many of the men to doze. Ten minutes later the vehicle came to an unexpected halt, jolting the men from their reverie and catapulting each roughly forward onto the man ahead. Heron reached out and inadvertently grabbed a haversack from the mound of gear beside him, which provided a handhold to help minimize the jolt. Each haversack in the pile held a raincoat, shelter-half, mess kit, toiletries, and cigarettes. And each blanket roll held cotton drawers, handkerchiefs, service shoes, socks, undershirts, blankets, clothing, half a two-man tent, and poles with ropes for assembling the tent halves. Loose items in the pile included cartridge belts, canteens, first-aid pouches, bayonets, and entrenching tools.

Waiting to board ship, June 1944.
U.S. National Archives Photograph courtesy of the World War II Database.

From behind the truck, Lt. Bonafin's voice cut through the stillness like thunder behind a bolt of lightning, "This is it, men. Welcome to Plymouth-Devon." Knees buckled as each man hit the ground under the heavy load of gear strapped to his back. The spot where they assembled would normally offer a pleasant view of the sound but on this day the blanket of fog rolled in from Dartmouth Bay obscured most of the scenery. An eerie quiet settled over the waterfront as the men queued up under heavy gear to parade like weary Jonahs into the LCT (landing craft, tank) that lay with its ramp open like the gaping jaw of a whale.

Once closed up, the LCT began ferrying the men out to the U.S.S. Bayfield anchored a short distance from shore. At the hallway point, the coxswain announced that the Battalion Commander and Commanding Officer of the 8th Infantry Regiment had already boarded the Bayfield, each with his respective landing team in tow. As Heron's craft drew closer to the gunmetal gray transport, he reached for the rope ladders hanging like braids over the sides of the ship and just as they rehearsed on dry land began climbing, only this time with considerably more difficulty under the weight of full battle gear and with the ship heaving over rough seas.

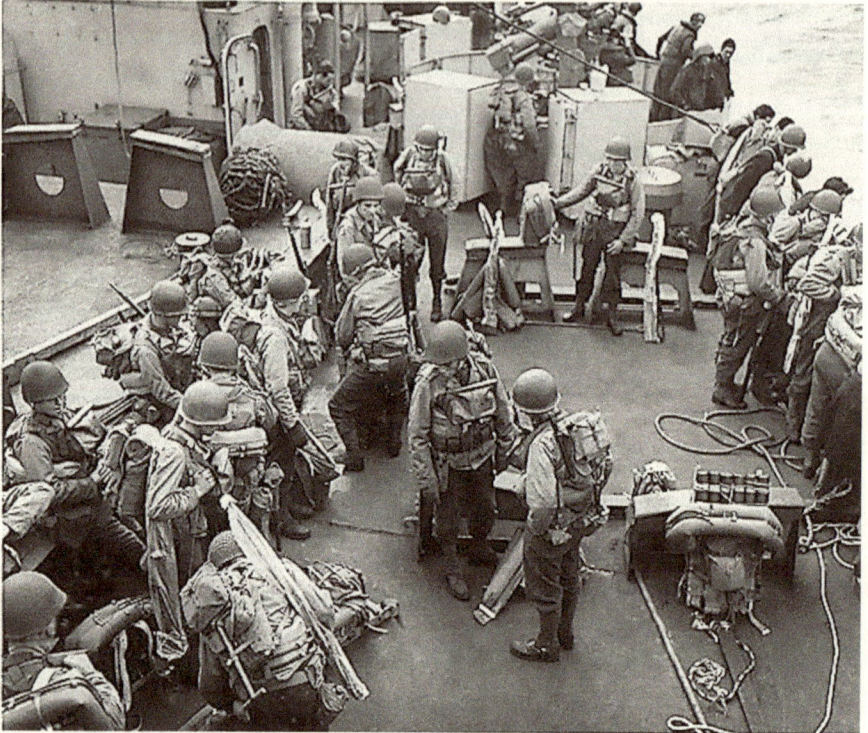

Under full gear.
US National Archives Photograph courtesy of the World War II Database

No sooner did Heron's gear hit the deck than the ship weighed anchor and stood out to sea. Four minutes later, Colonel Van Fleet came forward to announce the ship crossed the breakwater, adding that D-Day would commence on June 5th and H-hour at 0630 that morning. Heron sat quietly beside his gear holding a Garand in the crook of his arm with a thin plastic Pliofilm the men called an elephant condom, draped over the barrel to keep out the water. He could not imagine what fate held in store for him, as with each passing second the ship propelled him closer to a three-mile stretch of land codenamed Utah Beach located on the right flank, westernmost of the five Normandy landing sites.

After two days aboard the vessel, most men sat lost in thought or partially dozing from nervous fear and exhaustion. Word circulated that those going in on the first wave were expendable, chilling words coming back now to haunt each man.

Sgt. Heron sat wondering if he might never see Azelia again, when suddenly he felt the presence of someone standing before him. Looking up, he instantly recognized Brigadier General Teddy Roosevelt, Jr. and leapt to his feet to render a salute but the general waved it off. "At ease, son." He spoke softly. "What outfit you with?"

"Company A of the 87th Chemical Weapons Battalion, sir."

"I watched you men in practice. You're damn good with those mortars. We will be relying on your speed and mobility throughout the days ahead." In a voice calm and reassuring, he wished Heron luck then moved on to speak with another man, leaving him much more at ease than he felt scant moments ago.

The general wore a knit cap, walked with a limp and carried a cane, having sustained a leg wound in World War I that barely slowed him down but could not stop him from carrying on with his sworn duty. Despite a bad heart he begged General Raymond O. Barton to permit him to go ashore with the first wave so he could demonstrate to the troops would not feel alone or abandoned by the brass.

Moments later, Heron heard Roosevelt leading a group of men in the *Battle Hymn of the Republic* and felt a chill as he joined in, his eyes growing moist when they came to the phrase, "As He died to make men holy, let us die to make men free." When the singing ended, Sgt. Meyers came forward. "Here's your helmet, Larry." The sergeant had offered to camouflage it for him using a net garnished with strips of burlap. He left a horizontal white stripe on the back exposed, as required by regulations to identify Heron to his own men as a non-commissioned officer. By contrast, officer's helmets bore a four-inch vertical white stripe.

Seeing Meyers face so ashen, Heron's hand inadvertently went to his shirt pocket to feel for his own six-pack of Dramamine. Meyers had been popping motion sickness pills like candy, which apparently did not help. "Better go easy with those pills, Sergeant," Heron offered.

Meyers seemed lost as he staggered away looking like a man who just polished off a quart of bourbon, his hands searching his pockets for more Dramamine.

At 0415 hours on June 4, RAF meteorologist and Group Captain John Stagg warned General Eisenhower that the weather would prove too stormy for the planned June 5th assault. True to predictions, the weather rapidly increased in fury, hell-bent on driving the armada from the sea, so Ike postponed D-Day for 25 hours.

The delay left Heron more time to contemplate what may lie ahead and to wonder how he would react under real fire. For the moment, at least, he felt somewhat like he did just prior to the annual football game with Northbridge High, except with higher stakes and more to lose than a football game. Glancing about at the faces surrounding him, he wondered how many would be maimed or killed, and at the same time caught them looking back at him likely with the same thoughts crossing their minds.

I'll make it, just like I promised, Azelia.

He always kept his promises.

Just for you!

TWO HUNDRED forty-five minesweepers scanned a wide expanse from the Isle of Wight through the Channel to the transport anchor line off the French coast, where mines provided the greatest naval threat from the Germans. Behind them came the LCTs, some loaded with jeeps and others carrying tanks outfitted with floatation devices, the tanks scheduled to land at H-hour minus five minutes to support the infantry. Behind them came LCTs towing trailers filled with ammunition as they streamed forth like schools of dolphins toward the French coast.

Six battleships, twenty cruisers, and sixty-eight destroyers brought up the rear of the flotilla, as though driving the powerful armada toward its destination. The mission of the battleships was to take out heavy concentrations of German batteries that included a hundred-twenty guns ranging from 75 to 210 mm. One of those ships, the *Nevada,* had managed to get underway during the Japanese attack on Pearl Harbor despite damage incurred by an enemy torpedo and a fire amidships.

At two o'clock in the afternoon on June 6, high winds exceeding fifteen knots filled the channel with choppy waves, and after fifty-two hours of bucking in angry seas, Heron felt anxious to plant his feet on solid ground despite the enemy threat awaiting him. Ten miles from shore, he saw sudden flashes of ack-ack, and silhouettes of small boats groping against the waves. Then came the deafening *pow! pow! pow!* of the Navy's big guns, like the bowels of hell erupting. The concussive force from shockwaves that followed, smacked his body with such force as to jar loose his brains and rattle them around in his skull.

In the beginning, not much went as planned. The 82nd Airborne had yet to secure the town of Ste. Mère Eglise, where half the 507th and 508th Parachute Infantry Regiments landed in areas flooded by the Germans, an act that caused many to drown under the weight of heavy equipment. The Air Force's bombing of coastal defenses failed miserably due to a combination of overcast skies, bad luck and misjudgments. At 0330 an NCDU team (Naval Combat Demolition Unit), armed with a ton of explosives, timed their landing to fall within an hour and a half of low-tide to clear exposed obstacles from the beaches. The team that landed on Omaha beach took the worst casualties, losing half their 180 men, either killed or wounded.

Seventeen hundred feet from land, the Battalion Commander's party as well as the Regimental Landing Team together with the men of the 87th descended into landing crafts lined up alongside the ship. Just as he practiced in Operation Tiger on April 27, Heron swung over the side and labored down a cargo net. Only this time the sea tossed rougher, the sky ran darker, the ropes felt slipperier, and his hands trembled so much that he feared he would fall.

All around them came the sounds of roaring seas, engines revving, and officers shouting commands. Snow-capped waves swelled to lift the LCVP (landing craft, vehicle and personnel) half way to the ship's rail before plunging it deeply into the next trough. The men had been warned to keep their helmets unbuckled so that if they fell, the force of the brim hitting the water would not break their necks.

Just as Heron let go of the rope ladder to drop into the craft, it pitched wildly, throwing him against the gunwale and drenching him with seawater. Rubbing his hip, he adjusted his haversack and sought out a comfortable position in the standing-room-only vessel.

Waves chopped the rails, washing down the men closest to the sides and spilling onto the deck to consistently maintain an inch or more water underfoot. First one man, then another began to retch until every burp-bag got tossed overboard. The smell wafting to the pit of Heron's stomach sickened him but he fought back the nausea, determined not to follow the others.

To take his mind off the stench, he struck up a conversation with the coxswain, pleased to discover he originated from Medway, Massachusetts, a town adjacent to Milford. They promised to look one another up if both made it safely home. Heron admired the courage of the grim-faced coxswain as he guided their craft as close to shore as possible, managing to keep from ramming it so far aground that an exit and return to the ship would become impossible.

The thirty-six men on board Heron's craft began wading ashore with shells exploding and spewing geysers of muddy water high into the air beside them. The weight of both water and backpacks dragged men down, making swift progress impossible. Fortunately, the Germans failed to complete their fabrication of water barriers and fewer than expected mines exploded. The NCDU teams previously cleared paths and naval bombardment punched holes in defenses, which would save many lives by the end of the day. Even German shelling inadvertently assisted the landing by opening craters that served foot soldiers some welcomed cover.

Normandy Beach Landing.
US National Archives Photograph courtesy of the World War II Database

Running forward, Heron dove headlong into one, spinning around to face the rear just as Sgt. Volcjak came sailing through the air to land with a thump beside him. "Damn. My men are all over the place and we have no equipment," Volcjak shouted.

"What happened?"

"Mortars and ammunition went to the bottom of the channel with the LCA. My squad got picked up by another craft that dumped us ashore. Gotta find the lieutenant." With that, the sergeant climbed out of the crater and disappeared.

Heron moved fifty yards further along a causeway to collapse into another shell hole. Along the way he passed dead and wounded, most attended by medics. Another hundred yards brought him to a seawall where soldiers moved past a lively General Roosevelt wielding his cane at them as though directing traffic.

As history would note, the Utah landing took place a mile off course due to unexpected currents but instead of panicking when he learned of the error, the general kept his cool and uttered the immortalized words, "We'll start the war from right here." Like his father, the general would later receive the Medal of Honor for his display of courage, gallantry and leadership in the heat of battle, the highest of military medals bestowed by the President in the name of Congress. Thus they became the second father and son to win a Medal of Honor.

Teddy Roosevelt, Jr. stopped leading only when the piercing whine of a screaming low-level German ME 109 sweeping in to spit death and destruction, forced him to dive for cover with streams of bullets kicking up sand at his heels. Just as suddenly, bursts from a British Spitfire chewed a path across the German plane, breaking it into pieces to send it slamming into the sea.

With that threat removed, the General rose to his feet to continue waving men forward with movements normally restricted by a severe case of arthritis. Prior to the invasion, rumors circulated that he was stricken with pneumonia, which would sideline him from the battle. The ravages of his sickness clearly on display, he nevertheless kept moving forward shouting for his men to follow, as though energized by the gods of battle.

As Heron started forward, a deafening blast knocked him sideways and his helmet slammed a concrete piling. With ears ringing, Heron stopped to shake off the cobwebs and scan the area for signs of his squad, miraculously spotting all seven within a hundred feet of his position.

Branson lay prone a few feet to his right, talking into a radio. "Camel Green! Camel Green! This is Camel Red!" He turned to Heron and shouted, "Set up your mortars behind that tetrahedron." Then he was back on the radio. "Camel Green..."

Heron's Company A went by the code-name "Camel Red," battalion headquarters Camel Green, and companies B, C, and D Camel Purple, Blue, and Orange, respectively. The helmets of the men of the 87th bore a camel the size of a half-dollar stenciled on the left side in a color that matched their respective codenames.

Lieutenants Bonafin and Cable, Sergeants Burt and Faber, and Cpl. Trant acting as forward observers, came ashore at H-hour with the 8th Infantry Regiment of the 4th Division, in the wake of the NCDU teams.

Companies A and B arrived with the first wave, at H-hour plus 50 minutes in support of the 1st and 2nd Infantry Battalions respectively. When A Company assembled, Fiske ran out ahead and placed the distant aiming point 200 meters from the mortar and leveled the elevation bubbles while Sgt. Heron set the mortar sight deflection and elevation. After Shanahan added the boresight to the mortar and leveled its bubbles, Heron adjusted the elevation knob and set the deflection, ensuring the proper sight setting on the fixed deflection scale.

As if on cue, a Weasel (M-29 cargo carrier) pulled up beside them with 50 mortar rounds onboard, and less than a minute later the entire company began firing in response to directions from forward observers. Company's A and B fired 100 rounds and after twenty minutes moved inland to keep up with the advances of their respective infantry battalions. C and D arrived forty minutes later in support to the 3rd Infantry Battalion and the 22nd Infantry Regiment to fire another forty rounds before pulling up their mortars to charge forward.

For the next six hours, the 87th would provide the 4th Division its sole artillery support, firing at enemy strong points, machinegun nests, pillboxes, and concrete emplacements as directed by forward observers. The commander of the 4th Division would subsequently report the amazingly rapid and accurate firepower the mortars delivered whenever called upon. This marked the first time the 87th fired rounds over the heads of its own troops against an armed enemy.

German counterfire provided another battalion first, albeit an undesirable one. No sooner did Heron's squad spend its first forty rounds than the earth 50 yards from their position erupted into smoke and flame. The violent explosions continued to spray plumes of dirt on the men, shooting chunks of rock and shrapnel whizzing past their heads. Thunder and fury matching the power of multiple lightning bolts spewed earth and fire on all sides. A large piece of shrapnel whistled past Heron's ear as he helped dismantle one of the mortars then ran for cover carrying a tubular barrel cradled in his arms.

Enemy artillery batteries and mortar units that pounded the beach with heavy shells throughout the landing redirected their fire to begin honeycombing the town of Pointe du Hoc, signalling the failed plan of the Airborne Rangers to neutralize German artillery. Amid the din, Heron learned to distinguish the rattle of enemy machine guns from small arms fire, the pounding of 88s from the staccato of anti-aircraft guns, and to notice the differential sound of weapons the Germans captured from stores once belonging to the Allies.

Despite the enemy wall of resistance, the 87th suffered few casualties. A post-landing inventory showed the loss of two mortars and two vehicles along with a pair of LCVPs. Pfc. Smith of Company C was killed by a direct hit on his foxhole, and Lt. Cooper stepped on a mine and lay for a time in the middle of a minefield with black and red guck leaking from what remained of his lower extremities into the ground. The lieutenant simply looked up with glazed eyes when Fletcher reached his side. "Lieutenant," Fletcher cried out to him.

"I'm here, Fletch..." Though his eyes remained open, the lieutenant had gone.

Crammed aboard an LCT.
US National Archives Photograph courtesy of the World War II Database

Overdosed on Dramamine pills, Sgt. Red Meyers attacked the beach like a drunken sailor but when the fighting ended would not recall a single event he lived through. Of the 23,000 men who landed on Utah Beach, only 197 became casualties, a lot fewer than the 749 lost during training exercises in England, and far less than the 2,400 slaughtered on

Omaha Beach that same day.

On June 7, Company A supported the 8th Infantry Division during their advance on Ste. Mere Eglise and throughout the next twenty-four hours fired approximately 1500 rounds on strong points and enemy personnel, and by the end of the period convinced the infantry for all time of the superior accuracy and effectiveness of the hitherto under appreciated and relatively unknown mortar in the hands of highly trained soldiers.

Mobile Mortar unit delivering highly accurate firepower.
Signal Corps photo courtesy of the World War II Database

Battalion headquarters passed through Ste. Marie-du-Mont to dig in at Les Forge while Company A moved inland to support the 1st Infantry Battalion, 8th Regiment, setting up its mortars near La Madeleine. From a position concealed between hedgerows Heron's platoon passed the night deprived of sleep while high explosives and white phosphorous rained down on Les Forge from German units dug in at Turqueville and started a fire that by morning had completely destroyed the village.

The next day, Company A moved forward with the infantry, the 1st Platoon led by Lt. Branson and S/Sgt. Julian Brunt, the 2nd Platoon by Lt. Lesh and S/Sgt. David Thomas, and the 3rd Platoon by Lt. Bonafin and Sgt. Heron. By the morning of June 8, the company parted with the 8th Infantry Regiment to lend support to the 12th Infantry Regiment of the 4th

Division making an assault generally along the main road connecting Ste. Mere Eglise and Montebourg. Company A fired over 1000 rounds to help propel the 12th Infantry northward.

The company endlessly lent their support to one infantry unit after another in place of the rather cumbersome artillery, which works well in defensive or stationary situations but falls short when it comes to offensive charges requiring close-in direct support delivered in the blink of an eye.

On June 10, as the commander of Company D passed from the first platoon to the second, an enemy plane dropped a bomb that landed at his feet, killing him instantly. No one slept a wink during long hours devoted to taking out three German 88s and four machinegun nests with direct mortar hits. By morning the men felt so hungry they didn't mind eating from the "golden cans," a.k.a. the much-maligned Army C-rations.

By the morning of June 11, the 87th Chemical Weapons Battalion counted 24 casualties. With the final push to take Montebourg under-way, Company A took up a position north of Joganville, where late in the evening under heavy enemy bombardment, Cpl. Wilkevich death by enemy fire brought the total of 87th members killed in action to six, while a shrapnel wound incurred by 1st Lt. Leah raised the number wounded to twenty.

Early on June 12 Company A engaged in the assault on the town of Ozville by the 3rd Battalion of the 22 Infantry Regiment when powder charges in a mortar shell accidentally exploded, severely burning Pvt. James O'Donnell's hands, while shrapnel from the blast wounded Sgt. Harry Faber.

As the infantry stormed city after city, the 87th worked in tandem to deliver murderous firepower onto enemy positions, with mortar pla-toons detached and reattached daily from one outfit to the next, moving from one strange-sounding town like Quinville to yet another like Ste. Colombe, Orglandes, Hauteville-Bocage, Valgnes, Delasse, each assault advancing them nearer the city of Cherbourg and ever closer to Sgt. Law-rence J. Heron's appointment with destiny.

The second phase of Eisenhower's plan listed two major objec-tives, first to capture Cherbourg and second to amass sufficient forces and materials for a break out toward Germany. Cherbourg represented one of two major port cities, the other Le Havre, located at the tip of the Coten-tin Peninsula, and both ports played heavily into the massive buildup of personnel and supplies required to support twenty-nine combat divisions during the push toward Germany. Original plans called for the Allies to take Cherbourg by June 21, and pressure continued building toward making it happen as close to that date as possible.

ON JUNE 19, Heron turned twenty-four and that morning opened a letter from Azelia containing a string of black rosary beads with a solid silver crucifix included for his birthday. When he finished reading the letter he counted his prayers on the beads, kissed the crucifix, and slipped it into his left breast pocket.

Two days later, on June 21, Company A consolidated a position on the outskirts of the city in preparation for the attack on Cherbourg led by the VII Corps. A growing problem resulted when ammunition convoys from Utah Beach failed to keep up with the voracious demands of the infantry for artillery support, and as a result the infantry by now became totally dependent on the mortar men for their "artillery" needs. Not only could mortars provide extremely accurate fire power but the men could launch them with far greater mobility than their artillery counterparts.

On June 22, Company A dug in at a point northwest of Valognes to sidestep a 9th Air Force attack launched on Cherbourg's high ground then took up a position on a hill northwest of Delasse, closing three days later within 4000 yards of Cherbourg. On that day, General Omar Bradley ordered a naval bombardment on German batteries located in the city, and that same day three U.S. ships became locked in a three-hour duel with German gun batteries off Cherbourg's coast that ended with the loss of 52 U.S. sailors and damages to the battleship Texas and the cruiser Glasgow.

Just four more days remained before Heron's life would change forever.

CHAPTER EIGHT

SUICIDE MISSION

IN THE EARLY morning hours of June 26, Company A dug in behind the 39th Regiment in the vicinity of St. Sauveur-le-Vicomte, about 2000 yards from Cherbourg center. The VII Corp then moved forward to tightened its circle around Cherbourg as advance units of the 9th Division took up positions in the dockyard. During this latest advance, 1000 Germans fell prisoner, including garrison commander General Schoieken and Naval Commandant Admiral Hennecke, the man who earlier ordered the harbor destroyed to render it useless to the Allies, for which Hitler would award him the Iron Cross.

At this time, the 87th Battalion's ammunition dump took a direct hit from German artillery that destroyed its entire supply of mortar rounds, and replenishment would take hours, possibly days. By 1800 hours, Company A counted its last eighteen rounds and Captain Stiefel issued a desperate call for volunteers to retrieve hundreds of shells from the back of a disabled two and-a-half ton truck a hundred-fifty yards closer to enemy lines.

Sgt. Heron called for volunteers, but when no one would come forward answered his own call and by advancing alone toward the truck, despite pleas and warnings of snipers from Cpl. Bartosiewicz and Cpl. Madeiros. He darted from crater to crater unaware that a Waffen SS sniper, an elite member of Hitler's ablest assassins, kept crosshairs on him all along his perilous journey. He planned to load 50 rounds into the Weasel parked behind the truck and deliver them to where the men waited to begin firing at an enemy emplacement that at that very moment fired continuously from a location protected by a cement compound. While he carried out his lone mission, the German emplacement kept the trapped infantry battalion pinned down and marked for total annihilation. If he could safely bring in the first load, his men could come forward to help retrieve the rest of the ammunition from the truck.

Meanwhile, three thousand five hundred and forty-two miles away, Azelia awoke in her lonely bed with a start, her nightgown damp from cold sweat. Locked in a dream, she vaguely recalled Larry dressed in his army uniform wearing such a sad expression that she began to sob. When she awakened, she made her way down the hall to the bathroom to turn on the hot water and reached down to stop up the drain. While waiting for the tub to fill, she stared into a mirror above the sink at sunken eyes looking back with disconsolate terror. Returning to check on the tub, she watched the ripples spread to the sides and kneeled to swish water around to test the temperature. Then climbing into the tub and lowering herself into the water, she felt her muscles slacken and inexplicably the tears begin to fall.

AN ALARM SOUNDS at the nurse's station in the Milford Hospital, indicating a problem with the woman in Room 201, and within seconds the nurse arrives at Azelia's bedside to find a low-level reading on the IVAC machine. The nurse refills the bag then checks the monitor noting Azelia's blood pressure and pulse rate running on the high side but nothing to cause alarm. Her temperature reads a normal 97, so the nurse adjusts the covers and leaves the room.

Alone at last, Azelia looks straight ahead at the moon approaching the quarter mark outside her window. When the monitor sounded, she felt no fear of impending death, only annoyance because it interrupted her reverie.

Larry faced death many times in his lifetime, but the worst came on that day in June when fate delivered him to death's doorstep. On that fateful day he made the costliest decision of his life by acting to save the trapped infantry battalion.

At the same moment he faced life-threatening peril in France, she could have been laughing or enjoying a conversation with a friend, with absolutely no clue at the time of the horrors he faced thousands of miles from home. How could she? But not long thereafter, she would receive an ominous warning, not from the Army, but one that came from a psychic friend who foretold the worse news imaginable derived from a most unusual source.

CHAPTER NINE

OMINOUS PREDICTION

FRIDAY AFTERNOON found Azelia pacing her kitchen floor while awaiting the mailman's second arrival for the day, twice a day delivery the norm in 1944. Larry's last correspondence arrived more than three weeks ago, so it would make her day if she received one this afternoon.

She sat detached, quietly staring out a window past a fan droning on her kitchen table that billowing the white-lace curtains dancing gracefully before the window they adorned. Growing restless, she rose from the table and almost welcomed the sound of the phone that rang precisely when the Regulator clock struck two. Amelia, her sister, wanted to know if Azelia would like to accompany her and their sisters, Olga and Vera, on a shopping trip to Milford the next morning. After lunch they would go to the next door to drink Turkish coffee with Amelia's neighbor, Virginia Kalpagian, to have their fortunes read, an act she performed countless times with uncanny accuracy.

Why not? It would help take her mind off worrying about Larry, at least for a while.

At 2:30 PM the postman dropped off a copy of Life Magazine and the usual pile of bills, but nothing from Larry. "Perhaps his letters are queued in a central repository and will arrive in one huge bundle," the postman suggested. "Just give it a little more time."

The sisters arrived at Amelia's house near the other end of town on Dutcher Street precisely at noon the next day, where wonderful aromas greeted them the second the door opened. Virginia's husband worked on Saturdays, so she readily welcomed the opportunity to enjoy the company of good friends. She had known nothing but hard work from the time she grew up in "the old country" and possessed lots of natural energy. An Armenian who immigrated to America to escape the oppression in Syria, she quickly learned the language, worked as a seamstress for a while then opened and operated a children's clothing store in downtown Milford, where she designed and manufactured her own line of children's clothing in the store's back room.

From left: Olga, Vera, Azelia, Larry, Livia, and Guido.
Courtesy of Larry Heron, Jr.

A bright woman who learned to speak several languages and to play the violin as a child, she soon determined the town too small to support her business, so she sold it to competitors. At around this time, she began dividing twelve hours a day between stitching for a clothing manufacturer in Milford and stitching more clothes for them at home for extra money. Whenever the opportunity came to spend weekends cooking for friends and playing hostess, she seized upon it as a welcome means of relaxation and a break from her monotonous routine.

The women stepped into her house and before the door closed behind them, all stopped to listen to a low rumbling noise that quickly escalated into a roar. Moments later an army convoy pulled to a stop directly across the street, and an Army officer leapt from the lead jeep to begin shouting orders at disembarking soldiers. Squads of men entered the dense woods in the vacant acres across from them and dispersed to set up their equipment in orderly fashion. Soon fifty-caliber machineguns and small artillery pieces were set in place and concealed beneath camouflaged cloth, brush and broken branches. The men carried bazookas and rifles and took but moments to complete their exercise.

Watching the soldiers in action, Azelia hoped Larry would never see any fighting. The others found these maneuvers chilling but not unusual, for this was a country at war, and only forty miles separated Hopedale from Ft. Devens to the north. Virginia ushered her guests through the living room to a country kitchen in the rear of the house where they could look out windows through thin-laced curtains at a huge garden overflowing with vegetables. To help the war effort, nearly every family in America owned a "victory garden," this one spread over a full acre of rich soil.

"Coffee's ready," Virginia announced as a small brass ibrik frothed noisily on the stove, threatening to overflow. She poured a demitasse for each guest at the kitchen table where they sat around sipping the strong beverage and exchanged small talk. When finished drinking their coffee, each inverted her cup and set it down on a saucer to allow the fine grounds to assume a telltale shape. The women all grew quiet, turning their full attention on the one person who could make sense of the patterns that had taken shape at the bottom of each cup.

When she felt the proper time had elapsed for the dust-sized grains to settle, she reached for Amelia's cup but her neighbor raised a hand in protest. "Azelia first," she said. "She hasn't heard from Larry and is dying for some good news."

"News isn't always good," Virginia said, stolidly.

As a wisp of cool air sweeping down from Montreal gently raised the translucent white curtains, Virginia lifted Azelia's cup and turned it slowly in her hands, gazing at what to her practiced eyes represented a clear and meaningful view of unseen events.

Azelia leaned forward expectantly, hoping to hear that the war would end and Larry would soon return home.

Virginia's gaze slowly turned to a frown that broadened deeply and spread across her brows.

"What is it," Azelia asked. "Something about Larry?"

Virginia hesitated.

"What do you see?" she repeated, her heart racing and fingers absentmindedly tapping the edge of the table.

Virginia made eye contact over the rim of the cup. "He's in a place where people wear crowns." She hesitated, clearly holding something back.

"Crowns? Like kings?" Vera asked. "France? England?"

Virginia, suddenly seemed reticent. "A place where they wear crowns," she repeated.

"If it's England, then he must be okay," Olga said, with an encouraging glance at Azelia.

Virginia's face had turned ashen. "No. He's not... not okay I mean. He's..." Her search for the right words ended with, "It's not good." Then lowered the cup to the table and reached for Amelia's overturned cup.

"No, wait! What does it mean?" Azelia pleaded.

"Maybe he's just sick or something," Vera offered.

"It's worse," Azelia said, resignedly. Virginia said "not good" but the dire look on her face told more. "Tell me. What's happened to him?"

Finally Virginia relented. "He's in a building and he's not in good shape. I'm so sorry, but that's what I see." Virginia tried but failed to mask her horror. To make matters worse, her past predictions regarding Azelia's fortunes all came true with unsettling accuracy.

Despair swept through the kitchen like a dark cloud, eroding the bright and cheerful disposition Azelia entered with in an instant. Azelia became convinced in the moment that Virginia saw more than in the bottom of the cup than she cared to relate.

That night, Azelia told Emma, "Tomorrow, I'm going to apply for my driver's license." It made no sense to her until now because everything she needed, including her workplace, lay within a few short blocks of her home. She expected one day soon, Larry would return safely home to handle all the driving, but with the news she just heard from Virginia, she no longer felt sure. Her predictions had never proved wrong.

ON JULY 16TH she toweled off following a hot bath, added skin lotion to dry areas, threw on a sweatshirt and pair of slacks and trod barefoot down to the kitchen to start a pot of water for tea. In this kitchen she found solace, with its pale yellow walls, white moldings, a sturdy wooden drop-leaf table and bucolic pictures artfully framed and hung on the walls. Every article placed with an artful eye; nothing too extravagant or over-bearing and everything just right for its place.

She poured a steaming cup of the tea and took it to the kitchen table where she sat down to read the morning paper. Before taking a first sip, the doorbell rang with such urgency that even before opening the front door she felt certain bad news waited on the other side. When she swung it open, a boy she did not recognize stood before her holding a yellow en-velope, and he spoke the words every person in America dreaded to hear, "Telegram for Mrs. Heron."

Words formed a logjam in her throat. "I...I'm Mrs. Heron," she ut-tered mechanically.

"Sign here please." Her hand trembled as she scrawled out her name. *Please God! No!*

She would remember this day and everything she did from the mo-ment she set eyes on the boy with the telegram; her heart skipped and she suddenly felt faint. Lowering her body into a chair in front of the dining room table, she laid the telegram down but did not dare open or read it.

Emma suddenly appeared from the kitchen. "Aren't you going to open it?" she asked anxiously.

Azelia did not respond, for until she opened the letter and read its contents, only after she read the words, "killed in action," would he truly be gone. Until that moment, he remained alive.

"Maybe it's not bad news. Perhaps he's on his way home," Emma said. The terror in her eyes belied words uttered meekly.

As absurd as it seemed, Azelia felt that his living or dying depend-ed entirely on whether or not she read the telegram. Perhaps it would seem solipsistic or irrational to someone else, but it made warped sense at the time. "Go on," Emma urged. Then with shaking hands, Azelia tore open the envelope and removed the telegram but instead of reading, she thrust it forward. "Here," she said to Emma. "You read it." Then she closed her eyes and offered a silent prayer while listening.

Emma read aloud: "REGRET TO INFORM YOU..." Azelia's heart leapt. "HUSBAND SERGEANT LAWRENCE J HERON WAS..." She grew light-headed and found it difficult to breathe. "SERIOUSLY WOUNDED IN ACTION TWENTY SIXTH JUNE IN FRANCE." Emma dropped the telegram on the table in front of her and collapsed into a chair. Bad news fell on the women like a sudden downpour, drenching them in despair and a sudden wave of fatigue that weighed heavily not only on their bodies, but on their spirits as well. "At least he's alive," Emma sighed, at last.

"It said serious. How serious?"

"Any serious is bad enough."

Azelia heard Virginia's voice: *It's not good.* Her head felt light, thoughts spinning. *You can't die, Larry Heron. Don't leave me. Please dear God, you mustn't take him.*

"We must hope for the best," Emma declared.

Azelia laughed bitterly, without mirth. "And expect the worst, I suppose." Nothing in her well-ordered life made sense any longer. She let her head fall forward into her hands as if in utter defeat. "It's all so relative, isn't it? What makes sense and what doesn't, I mean." Yesterday she was smiling, happy, anticipating the day he'd return, and now...

"Don't worry," Emma said, unconvincingly. "He's strong. If anyone can make it, he will."

"Why didn't they tell us the extent of his injuries or where he is?"

Emma steeled herself. "All we can do is pray for the best and wait, dear."

Azelia clung to the hope that no matter the extent of the injuries, he would recover. Pick up any newspaper practically any day of the week over the last two years and there you would find articles about soldiers missing in action, injured, or killed. Though no consolation, she felt sure that thousands of similar telegrams went out every day. Larry was strong, a survivor. He would make it.

That night Azelia pushed the food around on her plate, tidied the kitchen and retired to her room early where she dragged out the photo albums. All photographs displayed him in black and white except for the large wedding picture that hung on the bedroom wall. Larry looked so handsome in all of them – so virile, so full of energy. That night, she cried herself into a fitful sleep.

She sat wistfully at breakfast the next day, a Tuesday, looking at her food and wanting to throw up. The thought occurred to call in sick but Emma convinced her it was better to keep busy. Later that night she called best friend Lt. Norma Ripanti, home on leave from the army. Norma concealed her shock. "It'll be all right," she said. You won't know the full extent of his injuries until you see him, so don't think the worst. I'm sure it's not as bad as it seems."

If only.

The Milford Daily News carried an old picture of Heron in uniform, commenting only that he'd been seriously wounded. The news sent shockwaves reverberating throughout New England, where Heron had carved his name as a sports legend with a promising career.

On her way home from the grocery store, Azelia paused before a wall in the center of town. An eagle crest painted across the center-top separated the words "TOWN OF HOPEDALE" from the words "ROLL OF HONOR." A row of flowers ran the full length of the honor roll wall that stood tall behind a low white picket fence. The wall contained more than three hundred fifty names with a star beside two names to signify the ultimate sacrifice, killed in action. One of the men on the list remained a prisoner on the island of Bataan, a place rapidly to become known as hell on earth for the 75,000 who surrendered to the Japanese, 12,000 Americans and 63,000 Filipinos. Many others returned home to lay in hospitals or rehabilitation centers, recuperating from wounds. The expression "war is hell" now held new meaning for Azelia.

Every town in America displayed an honor roll, and in the front window of nearly every home in every town hung a black star on a white background to proudly signify a household member served in a branch of the armed forces. A gold star meant something else that no family wanted to display, that the family member died serving his or her country. Whenever Azelia stopped to look over those names, she felt as she did entering a church; the feelings of reverence toward those protecting the freedoms of their loved ones, and of the nation struck her as overwhelmingly profound.

THE MOON had traveled one-third the distance across the window as the patient in Room 201 recalled the last time she and her loved one viewed it together. Late one night, not long before he left for war, they sat side-by-side on a park bench located about twenty feet from Hopedale Pond to view it together a final time. Because she had spent her entire life in Hopedale, it also marked the only town on earth where they looked up at it together. How fortunate was that? If she must live her entire life in just one town, she could do much worse than spend it in Hopedale.

HOPEDALE WAS an idyllic town with a rich and unique history. Elsewhere the sun may rise, but in Hopedale it steals from behind the hills with a whisper to gently bathe the valley with rays of love, hope, and peace.

In 1841, Universalist Reverend Adin Ballou and forty-five followers invested in communal stock to fund the purchase of a 600-acre farm on the Mill River, a parcel of land belonging to Milford and known at the time as the "Dale." Close friends Henry David Thoreau and William Lloyd Garrison shared Ballou's blueprint for communal living, as did Leo Tolstoy who predicted that one day the world would acknowledge Ballou as "one of the great benefactors of mankind."

But a series of financial and political disasters bankrupted the Dale, and in 1856 Ebenezer Draper, a machine shop owner, bought up all stock certificates and took possession of the entire town. Fortified with several key patents for loom parts, he and his brother George started a small business that grew over time into an industrial empire that led the world in the production of textile looms and related equipment.

In 1886, after a long and bitter struggle, the brothers forced a break from Milford and founded the town of Hopedale, and for decades the company flourished, as did Draper wealth. The Drapers applied a good deal of their fortune to developing "America's model company town," where residents never felt compelled to attend town meetings because they could depend on the Drapers to always act on their behalf.

The Drapers paid well and provided outstanding medical benefits to employees. They built award-winning homes, renting them to employees for minimal dollars each week and sent out company workmen to make repairs as needed. Draper craftsmen paid annual visits to spruce up each home at no charge, and with high-level financial support from the Drapers, Hopedale boasted the state's finest educational system. Recreational areas stood second to none and the town center held a full complement of award-winning civic buildings for town operations.

Since the early 1800s all company homes connected to sewer systems while Draper-made sprinkler systems protected the factory as well as every civic building in town, features unheard of in small towns prior to that time. The benefits of working for the Drapers read like a laundry list, for example, Draper-sponsored Fourth of July Field Days included tennis, racing, the long jump, logrolling contests, and other sporting events.

In 1892, Eben Summer Draper became lieutenant governor of the state, and beginning in 1909 served two one-year terms as governor of Massachusetts, and like every government leader throughout the history of mankind, he used his power to benefit Hopedale in every way possible. The Drapers and the town of Hopedale would continue to prosper throughout the length of the war, as did the town's residents.

Hopedale, Massachusetts Honor Roll.
Courtesy of the Milford Daily News

THE LOUD CLUNK of elevator doors closing outside Room 201 interrupted her journey back in time and seconds later Azelia slid back on her bed, propping herself on pillows, giving thanks to God for delivering her to an ideal town where she could enjoy the support of close relatives, good friends, and close neighbors, especially helpful and considerate through the duration the worse period of her life that followed close on the heels of the receipt of that first frightening telegram.

ON JULY 17, 1944, Azelia received a brief letter Heron had dictated to a nurse simply informing her he'd been hospitalized. Three days later a letter arrived from the War Department that read: Dear Mrs. Heron,

This is to confirm your husband, Sergeant Lawrence J. Heron, was seriously wounded in action on 26 June 1944, in France.

Theater commanders submit periodic reports of progress and accordingly you will be kept informed as these reports are received. Such reports must of necessity be brief and will not include information concerning the nature of his injuries.

I assure you our hospitalized soldiers are receiving the very best medical care and attention and it is hoped that a favorable report in his case will be received in the near future.

The letter gave Heron's temporary APO address and bore the signature of the adjutant general.

The phrase that struck her as comforting was "a favorable report."

So when the doorbell rang on August 11, she went willingly, expecting to find Vera who promised to drop off a stew pot Azelia lent for a baby shower the prior day. Instead, there stood the young man whose presence she came to deplore. "Telegram for you, Mrs. Heron," he said flatly. "Sign here please."

With heart pounding, she tore open the telegram and read, "MY THOUGHTS ARE WITH YOU ALL WELL AND SAFE LOVE AND KISSES=LAWRENCE HERON." Nothing about his injuries. She rushed to find Emma to share what both women regarded as very good news.

NO FAMILY was immune to the horrors of war. Too often newspapers carried articles and displayed pictures of area men killed, wounded, or missing in action. On August 19, 1944, Azelia read that Peter Saltonstall, the twenty-three year old son of Massachusetts Governor Leverett Saltonstall had been killed in action, and her heart went out to him and his family.

On the morning of August 23, 1944, Azelia sat at her kitchen table reading the Worcester Gazette, noting more articles on the front pages of death and injury to armed forces members serving in the war. Lowering the paper to the table, she turned to Emma, who entered the room with a basket filled with clothes needing ironing. "When do you think we'll get more details of his injuries or find out when he's coming home?"

"No news is good news," Emma simply replied then quickly lowered her head adding pensively, "I do hope he hasn't lost a limb or something." Her voice trailed off.

The next day, accompanied by Amelia, Azelia paid another visit to

Virginia's house on Dutcher Street, looking forward to an offer of Turkish coffee and another reading. When finished with her coffee, she turned the cup over to let it drain then handed it to Virginia, who this time seemed reluctant to take it. After the results of her last reading, Virginia did not care to repeat her role as the bearer of bad news. Azelia persuaded, "Just this one more time, please? You were the only one who had it right and besides, how much worse can it get?"

The ideal town of Hopedale provided a perfect setting.
Courtesy of Larry Heron, Jr.

Holding the cup so that the light showed the forms inside, Virginia turned it slowly in her hands, hesitated a moment then said, "I see long rows of buildings with a large flag flying from a tall pole out front. When you step inside, three gray ladies will come forward to greet you, each carrying a large bundle." She stared into the cup then added, "When you leave, you will take the three bundles with you."

The call came at six o'clock that evening, the woman on the other end identifying herself as a nun. "Mrs. Heron? I'm at the Municipal Airport in New York with… with a man who loves you very much."

"Oh, thank God. Is he all right?" Soon she would hold him and learn first hand the truth about his injuries. "Can I speak to him?"

"I'm afraid he can't come to the phone."

"Tell me, how badly was he injured?"

"You'll have to ask his doctor."

"When can I see him?"

"I don't know."

After a moment of silence, the nun said, "Like I told him, you really shouldn't see him this way."

A pencil fell from Azelia's hand onto the hardwood floor and rolled noisily to a stop.

The nun added, "But he told me that you will insist on it, that you love each other that much."

"Yes, sister. I do. I love him and I want to see him as soon as possible. Where are you taking him?"

"We're on our way to Framingham." The nun gave her the address of the Framingham Cushing Hospital, a 1,750-bed hospital just fifteen miles east of Hopedale. Azelia lifted another pencil from the holder and jotted the address on a notepad. "Many injured soldiers arrive there daily from the fighting overseas. We should arrive sometime late tonight and I am sure they'll let you in to see him tomorrow."

"Tomorrow then," Azelia said resignedly, knowing she would extract no more from the nun about his condition. Slowly, she lowered the receiver to its cradle while brushing away the tears. The telegram still lay before her on the table, right where she left it. Picking it up, she looked it over again then tried to piece together what she knew so far. Wounded in action then sent to a hospital in England, (a place where people wear crowns) with injuries so severe he could not come to the phone, not a good sign, as Virginia accurately predicted.

Azelia felt trapped in quicksand. The skimpy information given her to date did nothing to calm her frayed nerves. Perhaps tomorrow she would find the answers, but that night she did not get much sleep.

On Saturday she phoned the hospital early, hoping to find Larry settled in and the hospital staff willing to allow her in to see him. As with the nun, the nurse would not discuss his condition over the phone. Sergeant Heron needed rest after his cross-Atlantic trip and could receive no visitors until Monday. "Call back on Monday," the nurse told her.

At Ft. Rucker four months prior to Heron's life-changing mission.
Courtesy of Larry Heron, Jr.

CHAPTER TEN

THE PAINFUL TRUTH

ON MONDAY, Azelia spoke to a nurse then climbed into her car giving silent thanks that her driver's license arrived a week ago Friday. With Emma in the passenger seat beside her, she drove first to the Draper office building to pick up a ration book then up a steep hill to fill the tank at the Draper-owned gas station.

The "A" sticker on her windshield limited her driving quota to 150 miles each month, so she relinquished "A" coupons to the attendant for her four gallon weekly allotment. "B" coupons, issued to commuters with jobs essential to the war effort or located far from home, entitled the bearer to eight gallons while "C" coupons afforded doctors, ministers, farmers, and people directly connected with the war effort even higher limits. T-rations went to truckers and X granted unlimited supplies to those with the highest priority including ministers, police, firemen, civil defense workers and congressmen.

Gas rationing went into effect in May of 1942 after rubber fell in short supply. Since ninety percent of the nation's crude rubber originated in the Far East and most consumption took place on the roads, gas rationing seemed to make the most sense. Soon thereafter, Americans saw the list of rationed items expand to include sugar, meat, coffee, fuel oil, automobiles and similar items in short supply, such as shoes and typewriters.

After stopping to pick up Heron's sister, Ethel, Azelia turned the car onto Route 16 and headed east toward Framingham. Thanks to rationing very few cars passed them on the road. In addition to walking more, people stretched meals, planted victory gardens, and stayed home to listen to the radio.

When their car entered Framingham, Emma read from the directions Azelia wrote down before the trip. "Follow Winter Street to Mount Wayte Road, which should be around this next corner."

First a row of army barracks appeared then a large building with an American flag waving from a tall white pole out front.

I see long rows of buildings with a large flag flying from a tall pole out front.

"There it is!" said Emma. "That's the hospital."

Azelia felt grateful for the long drive because it kept her mind from cycling through the list of imaginable horrors lying in wait. As they drew closer, her heart lodged in her throat and her stomach performed repeated somersaults. She turned the vehicle left onto hospital property between Dudley Road and Winter Street and pulled into a parking space. Newly constructed in 1943 to care for wounded soldiers returning from the war, the hospital took its name from pioneer brain surgeon Harvey Cushing who served in the Medical Corps during World War I. Attached to the British Expeditionary Force in France, Colonel Cushing's greatest contribution resulted from his introduction of blood pressure measurement equipment in North America, the use of which spread through the western world like wildfire. Named the "father of Neurosurgery," Cushing passed away in 1939 at the age of 70.

When they entered the hospital, Ethel, pointed out three women seated on their right, exclaiming, "Oh my!" Simultaneously, Azelia spotted them seated behind a table the length of an aircraft carrier. Three nuns dressed in gray, nodding in unison and smiling at both Herons who quickly returned the favor.

Azelia led the way toward the reception desk adorned with a spray of roses and a bowl of pale out-of-season apples, where a neatly dressed young woman glanced up from behind a desk. "Can I help you?"

"We're here to see my husband, Sgt. Larry Heron."

"Ah, yes. We've been expecting you, Mrs. Heron."

The olive-skinned receptionist's delicate features and strikingly attractive face reminded her of movie actor Dorothy Lamour; she even wore her hair in a similar pompador style. "Please wait here a moment." Turning to her right, the receptionist motioned to the nuns who rose in unison to begin digging through two large boxes dragged out from under a table before them. Each nun lifted out a package then they came forward towards them like the procession of the Magi to begin stacking the packages on the receptionist's desk.

When you step inside, three gray ladies will come forward to greet you, each carrying a large bundle.

The closest introduced herself as Sister Elizabeth then the others as Sisters Mary and Alice. Gesturing toward their parcels, Sister Elizabeth said, "Volunteers knitted you these afghans to express appreciation."

"It's not nearly enough," added Sister Mary. "Just a small token of our nation's heartfelt gratitude to you and your husband. If we can ever do more, please let us know."

Sister Elizabeth smiled. "God be with you."

Each thanked the three nuns then watched them proceed down the corridor with gray gowns flowing as they rounded a corner into a long hallway before vanishing like ghostly apparitions.

In response to the question posed on Azelia's face, the receptionist said, "They are called the Gray Nuns, Sisters of Charity; volunteers in gray habits who nurse the sick, read and write letters for the war-wounded, sew on buttons, talk and listen."

"How thoughtful."

"You can leave your afghans here until you are ready to leave."

When you leave the building, you will take the three bundles home with you.

"Here comes the nurse to take you to see Sergeant Heron."

Azelia felt a rush like on her first day of school, anticipation mixed with fear of the unknown. "Please follow me," the nurse said. As they began walking down the same long hall where the nuns disappeared, she only half-listened to the nurse identifying herself as Frances Reid. A short stout woman with graying hair, heavily bagged brown eyes and skin as pale as the white outfit she wore, Frances led them out through a rear door to another building about a block away.

The procession entered and walked along a tomalley-green corridor on tomalley-green linoleum squares that gleamed under infinite layers of wax. They rounded a corner and entered another long hall made up of two sections: the first painted an institutional green and the second a light beige. Azelia wondered how a new hospital could look so old when they came to a stop before Room 107 C. She felt suddenly as though she was walking on air when Emma reached out to take her hand and the nurse grasped her arm. Azelia smiled ruefully at each of the women with eyes that failed to dispel the terror behind them.

The nurse pushed the door open slowly to expose a room dimly lit except for sunlight leaking in through partially closed blinds. A man dressed in white with a stethoscope dangling from a chest pocket stood on the opposite side of a bed looking down at a body wrapped like a mummy. Her knees buckled and she would have slumped to the floor but for hands reaching out to take her full weight as the strength drained from her like water sluicing down an opened drain. The women gradually lowered her into a chair beside the bed where the powerful smell of antiseptic flooded her nostrils causing her empty stomach to retch.

"What's going on?" Heron asked softly.

"Are you all right Mrs. Heron?" the nurse asked.

No! I'm not all right! She wanted to scream. "Yes. I...I'm fine." Azelia!

This could not be Larry, but the voice definitely was his.

What did they do to him?

Gauze covered his face except for an opening around his mouth. A tube dangled out from where his nose should have been through which she could hear labored breathing. The completely covered eyes added new meaning to the word shock and to her ever-augmenting fright. The scene reminded her of movies in which explorers gather around a mummy just removed from an Egyptian tomb. It could have been anyone under the layers of gauze, for his soft loving eyes did not gaze back at her.

She closed her own eyes and wished she would wake up in bed and find this a bad dream, but when she opened them again, nothing had changed.

"Oh Larry!" Gently she touched his arm, the only contact she dared, not knowing the full extent of his injuries. "Honey I love you so much. I'm so glad to have you home." The words sounded hollow even to her, like talking to someone through a door, not able to look into his eyes or read the expression on his face.

His mother said, "We're here, Larry. "Your sister and I will take care of you."

Emma's words cut into Azelia's heart like a serrated blade. Your sister and I? Had Emma interpreted her fainting spell as a sign of revulsion? Anger seized her as she choked back the tears and smiled up at

Emma, reproachfully. "I love you, Larry," she repeated. Calmer now, she leaned forward. "Everything will be all right. I will take care of you."

Cushing Hospital, Framingham, MA, circa 1951.
Courtesy of USGS

THE REUNION of husband and wife proceeded with all the makings of a wake. Azelia's mind never stopped whirling through space. When visiting hours ended, she said, "We'll be back tomorrow, darling. I love you." After everyone exited the room, leaving just the two of them alone, she said, "We're all so proud of you, Larry and everyone in town wants you to know that you're in their prayers." She felt at a loss as to what to say because inside she hurt deeply and felt miserable.

Azelia stood up and wiped at the tears rolling down her cheeks intending to plant a kiss, but with no place visible, she pulled back the covers, and to her horror found his chest wrapped in thick bandages as well. Lifting a sleeve, she located a clear spot of skin on his inner forearm to touch with her lips. When she looked up, she heard a knock as the door opened and the nurse who led her to the room earlier reappeared. "Sorry Mrs. Heron, but visiting hours ended long ago."

"Yes. Yes. I'll be back tomorrow, sweetheart." She kissed his arm once more then gently touched it with her hand. "I love you."

"Me too," he murmured, drowsily. Happiness, the excitement of her standing next to him, the lingering effects of a recent shot of morphine, all combined to consume what little energy remained, and soon after she left he drifted into a peaceful sleep.

"HOW MUCH WERE YOU TOLD?" Dr. Saenz asked the three women seated across the conference table from him.

"Nothing really," said Azelia. "The telegram indicated he'd been wounded in action but did not elaborate."

The doctor's eyes moved over her admiringly, for he saw a woman not only of great beauty but one in whom he sensed an inner strength and determination. Emma and Ethel leaned forward expectantly, leaving little doubt of their resolve to find out all that Heron endured. They would need all of their strength for what he planned to tell them.

When they departed the hospital an hour later, rain that started as a few large drops began splattering full force and they found themselves thoroughly drenched by the time they reached the car. But that did not matter. Nothing did but the heath and well-being of the man she left back there wrapped in gauze. Azelia even welcomed the chill that passed through her while slipping into the driver's seat as darkness fell; it fit her mood perfectly.

On the drive home, she could not shake the sinking feeling nestling into the pit of her stomach. Besides the sound of rain hitting the roof, the only other sound came from the wiper blades marking time like a metronome, not quite cleaning the glass and leaving streaks that created a distorted rendering of the outside world.

As she steered along Route 16, her thoughts journeyed to their now distant past to scenes they would never again repeat, because of his transformation. And though aware the doctors and staff held back details, she caught enough of a glimpse at the road ahead for them to see it paved with turmoil and tribulation. Clearly, to return tomorrow to face more bad news would require more strength and determination than anything she ever faced in her past.

Mesmerized by the multicolored blurs from street lamps crossing her windshield, she focused her attention on the muffled drumbeat of the wiper blades and the water from passing cars cascading over the hood in its rush to find gutters along the side of the road. Occasionally, she slowed to avoid splashing water on umbrella-toting pedestrians moving like vague shadows along the sidewalks.

In the rearview mirror, she saw Ethel leaning in one corner fast asleep, and at Emma with eyes closed and head resting against one window. Emma had proven a rock, the one person who truly shared her torment. Perhaps with luck and determination, they might find a way to restore some of the quality stolen from Larry's shattered life, and whatever joy this terrible war tore asunder. But judging from what she witnessed on this long day, she knew that the quality of both their lives as it existed before the tragedy, had deserted them forever. Life would never be the same, at least not without divine intervention.

HERON AWOKE six hours after Azelia's departure lying rock-still on his back with his mind alert to sounds and smells but his eyes gone to a dark new world. Except for the clock chiming three times somewhere down the hall, everything suddenly fell pin-drop still. Three hours after midnight and he would not asleep again until after his next morphine shot, five hours from now.

He tried forming a mental image of Azelia but failed and the horror of his dilemma sunk deeper still leaving him with a feeling he never before experienced in his life, a feeling of vulnerability on the verge of sheer panic. Lawrence J. Heron died out there on the battlefield, at least the Heron she knew and life as he once knew it. His wife stood by his side and he couldn't even look in her eyes or kiss her lips. Thank God he could still recall the luminous clarity in those trusting eyes that saw equal trust in the eyes of everyone she met. He took a deep breath, exhaling slowly to calm his frayed nerves. He must find a way to bring her face back into focus, to somehow recall the face that launched joy in his heart each time he gazed her way.

Images of her face flashed in his mind one small fragment at a time, her lips then her eyes, her hair perhaps her smile, but he could not hold onto them even for a fraction of a second. Finally he recalled the black and white picture he carried into battle with him and her face flashed black and white in his mind, but only for an instant. He tried replaying the wedding picture that hung on the wall of their bedroom in full color. This time he saw her clearly and he held onto that image for as long as he could.

"SINCE YOU'RE still awake, let's have another look at your blood pressure, Mrs. Heron."

"Let me know if you find any," Azelia jokes.

The nurse pumps air into the pressure sleeve then allows it to bleed off. "Pressure still looks okay. Sorry, but I have to check your insulin."

"Go right ahead. I feel like a human pin cushion the way you people keep taking my blood. Insulin shots I hardly feel."

"Insulin's low. I have to give you a shot and then will leave you alone."

"No problem." She really doesn't mind the insulin shots. The drawing of blood from her arms bothers her more. Last night they tried three times before locating a vein that would pump out enough blood. The only reason they take my blood, it seems, is because I am trapped here and they can. She knows that if she went home tomorrow the blood-letting would stop, so what's the point?

After the nurse leaves, she lies wondering how she reached eighty years of age so quickly. *Wasn't it just a few short years ago that I was young and energetic? My God, I'm an old lady, an old lady! My children are all grown and I have grandchildren in college and my Larry, he's been gone what, five years now?*

Bits and pieces of the horrors her husband endured on the battlefield found their way to her over a period of years from various sources. Larry rarely spoke on the subject of his ordeal, perhaps because the endless operations throughout his life served as reminder enough.

During the time she worked at the library to supplement their income, Azelia borrowed every book about the war she could find time to read, supplementing what she heard from returning veterans who fought alongside Larry in France, including two friends he knew before the war. Fr. Edward T. Connors came across Larry on the battlefield at the time he received his wounds and Dr. Joseph E. Murray crossed paths with him upon his return to the states, both meetings quite by accident.

And then there was this fellow soldier, John Sears, of the Fighting 87th, God bless his heart. John dedicated his life's work to chronicling the history of the 87th, supplementing it with daily journals and all the materials he collected from various sources from the time he served and throughout the remainder of his life.

Many years would pass before she would finally piece together the exact details of her husband's brave actions on the fateful day he saved an infantry battalion at the cost of his future, and of nearly everything meaningful in his own life.

With an impending sense of urgency, she strains to see where the moon disappeared, as it slowly creeps out from behind a darkened cloud. She must not waste precious moments if she hopes to complete the span of their lives before the moon slips away and morning arrives, the time when she suspects her life will also come to an end.

One Christmas, five-year old Azelia Noferi found a box with

holes cut into it waiting under the Christmas tree for her to open. The tiny paw that poked out through one hole belonged to a Siamese kitten that promptly ran up the Christmas tree upon its release from the box. The tree swayed beneath its weight, hovered for a second, then toppled with a crash onto the floor. As time passed, she bonded with the cat and learned that Siamese can be very protective. She smiles now, as she recalls the day it attacked a dog that came up her front walk sniffing a path toward the front door. When the dog spotted the Siamese, he pranced right up to test if they could become friends. The Siamese, full grown by that time, reared back and raked it across the nose to send it howling down the street.

The cat she named "Monkey" loved her so much that it would nuzzle her nose, curl up against her and blink its eyes at her to express deep love, not an easy thing for a cat. Each day when she came home from school, Monkey would come bobbing out to greet her, except the day she found him curled up in a corner of his little bed, too weak to move. It happened late at night and her father promised to take the cat to a vet early the next morning. That night she tossed in troubled sleep then watched the moon out her window until it disappeared to signal the arrival of morning. She bolted downstairs to look for the cat but sometime during the night, with its full moon, Monkey had departed, leaving behind a stiff replica of the Monkey she knew, and now the strong feeling persisted that tonight's moon promised she would meet a similar fate.

Therefore, she must hurry and finish recalling their journey through life, before the fading moon passes completely out of view, and her with it. With that in mind, she returned to that fateful morning, June 26, 1944 to relive the horrible scene played out on the outskirts of Cherbourg, France.

CHAPTER ELEVEN

HANDSHAKE WITH DEATH

THE COMMANDER of the Third Infantry Battalion yanked the SCR-300 handset from the radioman's bloody hand and screamed into the mouthpiece, "Camel Red! Camel Red! Did you get that?"

Lt. Eugenio L. Bonafin could barely make out the commander's voice above the cacophony of exploding enemy artillery shells when he yelled, "Yeah. I heard."

"For Christ's sake where's our support?" the Colonel cried. "We're pinned down by fire from behind a concrete emplacement and taking extreme casualties."

The lieutenant's eyes earlier took a scan of his men crouched down in holes dug to avoid shrapnel or a sniper's bullet; he commanded one of two platoons comprising Company A of the 87th Chemical Weapons Battalion. One of those men, David Rubenstein, held a paper and pen, describing their current location in a letter to his mom as a slaughterhouse.

Right now, the lieutenant focused his attention on Sgt. Heron, who landed in the shell hole from out of the blue, and immediately showed empty hands to signify the company had run completely out of ammunition.

Bonafin repeated the unthinkable into his handset, "We're out of ammo, Colonel. Damn it! Call for artillery!"

"All units from Cherbourg to Utah Beach are engaged; if one broke free this second it would still take hours to move within range and set up."

Bonafin cradled the handset and motioned Sgt. Heron closer. "We need a volunteer, just one man to load the weasel and bring it here."

The slang term "Weasel" stood for the M-29, a tractor-tread vehicle the size of a small tank. Manufactured by the Studebaker Corporation, the weasel could transport close to fifty shells at one time, a veritable workhorse that could maneuver over the roughest terrain imaginable.

"If one man brings in a load without incident, we'll know it's safe to send the platoon for the rest."

Heron understood a suicide mission when he heard one.

A half-hour prior, the truck loaded with ammunition hit a mine that took it out of action. The ammo dump which was its destination then took a direct hit and replenishment of all those lost mortars might take days. Although the mine blew the truck's engine, popped the hood and flattened the front tires, a hundred crates containing two twenty-five pound mortar shells each sat untouched and unexploded in its rear. If Company A's mortar-men could get their hands on the rounds on board that truck, an equal mix of white phosphorous and high explosive rounds, they could begin dropping fire on the concrete emplacement from their current position and take it out with pin-point accuracy within minutes.

"Get me a volunteer."

Sgt. Heron moved from one foxhole to another relaying the lieutenant's message to members of his squad calling for just one volunteer, but each man averted his eyes like a school kid hoping the teacher wouldn't single him out. "It's suicide," Cpl. Bartosiewicz said, perspiration beading his mud-streaked face. "Snipers are waiting to pick off anyone who shows his head."

At that moment, Heron made the costliest decision of his life. "Cpl. Madeiros, get on the radio. Tell Lt. Bonafin I'm on my way and when he gives you the signal, bring up the rest of the squad."

Turning to the company's top sharp-shooter, he said, "Cpl. Fiske, fire at anything that moves," then added, "except me."

Cpl. Bartosiewicz grabbed Heron's shoulder. "But sergeant, the snipers!"

Prior to joining the Army, Heron established himself as one of the fastest running backs ever to play high school football, unofficially clocking the forty-yard dash in 4.28 seconds; he set records for stolen bases in high school and led American Legion baseball in home runs. He repeated his record-setting ways for a semi-professional baseball team sponsored by the Draper Corporation, which meant nothing more to him, or to any of the others hugging their muddy shell holes on the outskirts of Cherbourg, France on this particular afternoon, other than he was the fastest amongst them, and one of the strongest.

Whoosh! A single artillery shell sailed overhead exploding a hundred yards to their rear and shaking the earth to its core. Each man mindlessly assumed the fetal position, knowing he would never hear the one with his name. Since landing on French soil two weeks ago, they learned quickly that a shell travels faster than the sound of the weapon that fired it, and "woosh" meant one just passed overhead, while the one that would "get you" would make no sound, no bang, no whizz or whoosh, just puff, and you're gone forever!

Pvt. Williamson leaned forward, his face caked with sweat and grime. "If you go, Sergeant, I go." The terror flashing in his eyes belied brave words but Heron took it as a noble gesture nonetheless.

"You don't have to do this." Madeiros pleaded. "You're the Sergeant." But those who stood by watching him prepare for action viewed Sergeant Heron as the only man with even the remotest chance of retrieving the shells.

"If there's a sniper, I'll draw him out," he called over a shoulder. Keep your eyes open and give me as much cover as possible." With that, Heron swung into action with intensity demonstrated again and again on the football field, only this time the field lay uneven, offering no protection, and the penalty for error spelled death.

The members of his squad watched Heron move forward twenty yards, dive for cover, pop back into view then dash twenty more, with his legs feeling more and more each time like they were turning to lead.

In the distance another pair of eyes followed his every move with strange admiration, holding the moving figure as close to the crosshairs as humanly possible. He watched as Heron pried the lid off one crate, remove a shell then a second, laying both rounds gently on the bed. Then the soldier leapt to the ground to carry the shells one at a time to a spot out of sight behind the truck, where he could not see him loading them into an M-29 shielded from his view.

While the rest of Heron's men and Lt. Bonafin watched, Fiske passed the sights of the Browning directly over the farmhouse detritus, ready to sweep it with .30 caliber bullets at the slightest movement or glimmer of a weapon.

Heron repeated the exact routine of dragging a crate to the rear, unloading first one then another, leaping to the ground, gathering strength then finally taking it to the M-29 behind the truck. The sniper wanted to know why he stacked them there but felt in no hurry to find out. Perhaps a second truck would pull up loaded with soldiers to help retrieve the rounds or to begin firing from that location. In either case the arrival of more men would afford the opportunity to take out multiple targets.

Sweat poured from Heron's body from sheer physical exertion and extreme dread. He had seen bodies jerk from the impact of a sniper's bullet even before the sound of the exploding cartridge reached the dead man's ears. He too could be dead in the next instant and never hear it coming.

German sniper holding a K98 Mauser with signature grenade launcher.
German Federal Archive photo courtesy of Wikipedia Commons

Little did he know that a sniper lay guiding crosshairs from his chest to his head and back all the while he followed his routine. Every time he jumped to the ground, Heron would turn to pick up first one of the two rounds resting on the truck bed, gather his strength then carry it to the hidden location behind the truck and then repeat the sequence with the second round.

With twenty crates thus removed from the truck, the sniper decided he should wait no longer and poised his finger a millimeter from the trigger so that the slightest caress would dispatch a bullet spiraling through the air to blow out the top of the target's skull, but he stopped when his eye caught new movement. Easing his finger off the trigger, he watched more soldiers coming forward to join the target.

Perfect! Now let them cluster behind the truck and go for a shell shot. Just a moment longer...

HERON STOPPED to wipe beads of sweat forming on his face like condensation on a cold bottle of Coke on a hot summer's afternoon. As he reached down to remove yet another round from its crate, he thought how much he could use a cold bottle at that moment.

The sniper observed six men closing in to lend a hand and steadied his finger, bringing it in contact with the trigger as Heron stood alone at the rear of the truck lifting the second shell from the current crate. The Wassen SS sniper centered his crosshairs, drew a deep breath and exhaled half then squeezed. The bullet followed an unobstructed path, hitting the shell in Heron's hands squarely, and unleashing the fury of seven pounds of chemicals inside the twenty-five pound shell, while simultaneously setting off a second shell at his feet. The combined intensity of the blasts dispersed shell fragments and lifted Heron bodily, tossing him twenty feet through space to land on his back with white smoke billowing from his flesh and clothing.

Diving headfirst into a blast furnace with thousands of volts of electricity passing through his body could not feel worse. The excruciating pain felt palpable, so intense that he took no notice of the blood pouring from multiple shrapnel wounds. But he did feel the phosphorous eating into his flesh like thousands of exploding match sticks and hoped he had instinctively closed his eyes in time to save them. With a surreal wakening, he knew that the voice he heard screaming belonged to him.

At the precise moment the bullet began its trajectory, Fiske spotted the muzzle flash and immediately began raking the pile of debris with bursts of .30 caliber bullets that shattered glass, ripped wood into fragments and turned chunks of stone to dust. The first burst removed the top of the sniper's skull, killing him even before a second entered his left shoulder leaving a gaping exit hole behind and below the right armpit. Fiske emptied his clip, inserted a second twenty-round clip and emptied it into the pile as well.

BARELY CONSCIOUS, Heron's senses approached shutdown, the

point where all cognitive thoughts cease and defense mechanisms kick in. With his brain on autopilot, nerve impulses ricocheted from one neuron to another, bouncing around in his skull like a flock of startled pigeons as he clawed viscerally at his face where fiery nuggets of white phosphorous burned deep craters into his flesh, eating into it. This was what it like to stick his face into a bed of hot coals and leave it there. No merciful unconsciousness ensued, no single glimmer of hope that the nightmare would fade, only an unfathomable indescribable pain.

Divine providence had deserted him, his lungs pumped like organ bellows attempting to re-inflate after the force of the concussion subsided, and his heart clunked in his chest like an off-balance washer on spin cycle.

Cpl. James N. Madeiros, Cpl. Edmond S. Bartosiewicz, Pvt. Michael A. Valette, Pvt. Arthur Almeida, and Pvt. Robert L. Williamson, stunned by the blast and knocked to the ground by its impact, slowly rose to their feet one-by-one, none with more than minor burns and superficial wounds. Williamson felt something foreign stuck between his lips and spit it into his hand, and to his horror recognized part of an ear, instantly flinging it away.

After the shock of the explosion abated, all focused their attention on Heron. No one expected to find him alive. Blood soaked his clothing and his face, what was left of it turned their stomachs. Each man, one after another looked down then moved quickly away. Madeiros ran several yards to where he leaned over and vomited.

An ordinary man would have died instantly, or so they thought, which might explain why each left his side. Every man present moved away convinced of his fate, felt powerless and could not watch. No one could survive one shell exploding in his face, let alone two, not even Larry Heron. But although he could not see them, he heard their voices as they reassembled, gathering a few feet from where he lay until he heard someone say, "We'd be doing him a favor."

Dear God! They're drawing straws to see who puts a bullet in my head. Never in his wildest dreams did he imagine it could end like this, always so sure he could survive any catastrophe no matter how grim. But upon hearing those words, the notion of his own mortality seized him. "No!" he cried. "My wife! She'll take care of me!" But the raw flesh of his mouth stuck dry and the words adhered to the back of his throat. *They don't hear me. Why can't they hear me? Please God. Azelia...*

Fiske jabbed at the pile of debris then the sniper's body with his rifle like a cat poking a dead mouse. The one eye remaining in the German's head stared up at him as a fly crawled out of the mouth, wiggled its wings then slipped back inside to lay its eggs.

THE MEN WERE drawing straws when Lt. Bonafin rushed forward, and they moved aside to open a path for him. "Hold on," the lieutenant said, dropping to his knees to feel for a pulse. "Jesus Christ, he's still alive." Burned beyond recognition, the lieutenant noted that both eye sockets welled with dark blood, most of his nose and both ears were missing, and his skin smoldered like barbecue ribs resting on burning coals. Bonafin looked at the blood from multiple wounds soaking into the ground and yelled, "For Christ's sake get a medic over here!" Then he tore open a pouch and removed a fresh compressor to replace the blood-soaked gauze Williamson held pressed to a wound in Heron's chest.

"Medic!" he shouted.

Convinced that the chemicals ate through his nerve endings, Heron could no longer feel his face.

"Medic! Get a fucking Medic over here!"

Bonafin could hear the distant sounds of fierce battle, signs that the trapped infantry still clung to life, badly in need of life-support. "Madeiros," he ordered, "have the men move these shells to the mortars and order them to commence firing."

"Yes sir!" Ignoring the ringing in his ears and the queasiness in his stomach, Madeiros rounded up the remainder of the squad. Minutes later Company A began delivering punishing fire on enemy positions and laying down a barrage to help the infantry break free. At least now, hundreds of men would survive because of Heron's courageous sacrifice, though it did little to ease their minds.

Cpl. Fiske rushed forward to address the lieutenant. "I got him, sir." He said excitedly. "I found the body of a German sniper in the remains of the farmhouse."

Bonafin cocked an eye. "Fucking snipers." As if seeing Fiske for the first time, he reached over and patted him gently on the back. "Good work, Fiske. You sure he's dead?"

NEARLY AN HOUR elapsed from the time of the explosion to the arrival of a medic who took one look at Heron then turned toward the lieutenant, "There's nothing I can do for this man, sir. Better I help those who can survive." He rose to his feet and began moving away.

"Get back over here and do whatever you can for this man," Bonafin barked with asperity. "He just saved an entire infantry battalion! And if you know what's good for you, you'll keep him alive!" The irrationality of his own words struck Bonafin but he didn't care because it did the trick.

The medic fell by Heron's side and took a longer look. The nose, ears and face were unrecognizable. After blotting his sockets with gauze pads, he found lids of torn lace over bloody pulp for eyes. Upon closer examination, he determined the left ear was missing and only portions of the right remained. Pieces of the forehead, lips, cheeks, and jaw were gone and blood oozed from shrapnel wounds like ketchup through a caldron.

The medic went to work, dosing Heron's wounds with water from his own canteen. "Lieutenant, hand me your canteen and tell your men to hand me theirs. Hurry!" He threw water on the burned areas and wiped away as much of the phosphorous as he could.

Then he applied sulfa and compresses to the wounds and hooked up an IV. He could not stop from thinking that had someone acted right away, had the men poured water on his face immediately, they just might have caught the phosphorous before it ate through his lids, possibly keeping it from burning deeper, destroying his eyes.

"What else can you do for him?" Bonafin asked.

"Everything I can think of," the medic responded. "He's not going to make it anyway." He groped Heron's pack for morphine, tugged on a sleeve then pierced his arm with a needle the patient did not feel. Normally, he'd have used Heron's blood to mark an "M" on his forehead to prevent someone from inadvertently giving him a second morphine shot that would prove lethal, but the badly burned and bloodied forehead prevented it, so he wrapped the top of his head with gauze then marked it with a large "M." He stared in wonder over how Heron remained alive, still awake even, and with his lips moving as though trying to speak.

"What's he saying?" Bonafin asked.

"I don't know. He's incoherent." The medic followed prescribed procedures by scraping at the raw and bleeding phosphorous burns with a brush from his pack while he poured more water on the burned areas. But he had never done anything like that to a raw face before and found it difficult. Surprisingly, Heron did not cry out in pain as the medic worked on the burned areas. Finished with that chore, he rose to his feet looking around for a stream or large body of water where he might soak Heron's entire body and really flush the burned areas but no body of water existed for as far as the eye could see.

Now he unsnapped Heron's canteen, unscrewed the top, removed a roll of gauze from his backpack and soaked it with water. Then he lifted the injured man's head gently and began wrapping the raw burn wounds that covered his face. This would help smother the chemical, cutting off the supply of oxygen. "Someone should have done this sooner but thank heavens most of the particles seem to have burned themselves out."

When finished wrapping Heron's face to resemble the head of an Egyptian mummy, he examined his dog tags to check Heron's religious preference. Rising to his feet, he shook his head and told Bonafin, "It's no use, sir. Better find a Catholic chaplain to read him last rites before it's too late."

Bonafin could hear the mortars coughing in the background as he mouthed a silent prayer then called for his radioman.

TWENTY MINUTES after the call went out for a chaplain, a jeep pulled to a stop about forty yards from where Heron lay dying. Painted on the sides of the jeep were the words *CONNORS COFFEE SHOP.* A soldier with a chaplain's cross painted on his helmet stepped briskly from the jeep and rushed to Heron's side.

Whether the dying man was Catholic or not made absolutely no difference to Fr. Edward T. Connors, the priest who had volunteered to join the fighting. A friend to enlisted men and officers alike, he had already attained the status of living legend and war hero. Men of every faith in the 9th Division referred to him as the "Great Chaplain," the "Soldier's Chaplain," and "Our Chaplain."

Connors Coffee Shop had become a 9th Division tradition. Everyone in the outfit knew they could always find a cauldron of coffee sitting atop a kerosene-heated ration-can in his tent. And whenever it ran low, the Chaplain would simply add more coffee, water or whatever it took to generate more moderately drinkable coffee.

Connors fell to his knees beside the wounded man, witness to more battlefield deaths than any one man should in a lifetime. Death by now struck him as much a part of the landscape as trees, rocks, farmyards, fields, and streams. He performed last rites on men who stepped on landmines or fell on grenades, men with bodies ripped apart with no chance of survival.

The Chaplain leaned forward to examine the dying man's dog tags then jerked back in horror as he read the name. "Dearest God!"

"What?" Bonafin asked. "What is it Father?"

"I know this man," said Fr. Connors. "I...I can't believe it's him! Larry Heron."

"You know him from somewhere?"

"I was the athletic director of a school in Fitchburg, Massachusetts when he played for St. Mary's, and he always beat us. He's the fastest and most powerful running back I've seen play the game."

"Larry." He waited for a response. "Larry Heron. Can you hear me?"

Heron's head rolled from side to side as his lips moved but the words came out as indistinguishable mutterings.

"It's Father Connors. Ed Connors from Fitchburg, Massachusetts. St. Bernard's."

Connors did not feel certain that Heron could recognize his name. Better perhaps if he remained unaware that a chaplain stood by to read his last rites. Gently patting Heron's shoulder to calm him, he slowly, hesitantly, produced a bottle filled with the "oil of the sick," wet his hand and began anointing: "In nominee Pa tris, et Fi lii, et Spiritus Sancti." In the name of the Father, and of the Son, and of the Holy Spirit. He followed a field form of the ritual aimed at destroying any power the devil held over a person awaiting death then went through the motions of anointing his eyes, ears, nose, lips, hands, and feet in that order, each time praying forgiveness for their wrongful use.

When finished, Fr. Connors slowly rose to his feet and stood looking down at Larry, his eyes brimmed with tears. Not much left to do now but wait for his young friend to embrace death, meanwhile he would continue praying that Heron's soul would find the right door.

CHAPTER TWELVE

BLACK FOURTH OF JULY

LATE IN the evening of June 26, a jeep rigged with four stretchers attached to metal brackets delivered Heron and three additional wounded to a clearing beside a temporary hospital. Each of the injured lay under a blanket to prevent the shock that accompanies blood loss and severe burns. Attendants carried the three who accompanied Heron inside the hospital tent where the lusty smells and noises peculiar to hospital surroundings greeted them.

Heron's stretcher remained behind on the ground outside with him still in it. The battlefield "triage process" called for the dying to be left outside because of the shortage of beds, medicines, and hospital staff. Studies proved over time that patients burned as badly as Heron usually survived for periods measured in minutes, suggesting the miracle would soon end and he would never see morning.

Heron kept slipping in and out of consciousness on the debilitating trip to the hospital. Fortunately, an orderly kept the presence of mind to continuously administer fluids as the least measure he could take for the dying hero. Despite the administrations, Heron's body experienced rapid dehydration and that condition together with the resultant lack of saliva, rendered him unable to speak.

The fact that Heron still showed signs of life two days after the jeep delivered him to the hospital remained a marvel attributable to nothing more than superb strength and pure stamina. The attending orderly who checked on him the next morning observed the patient seemed a bit more cognizant, so he spoke to a doctor about how Heron's condition had improved. The doctor then ordered the patient carried inside and placed on one of three beds that became available during the night. Even for the sighted, the dimly lit tent seemed a place of dark shadows packed with the severely wounded, an environment commensurate with Heron's blackened and obscure world.

Jeep rigged to carry four stretchers.
US Army Signal Corp photo courtesy of the World War II Database

It seemed that the longer he dealt with the loss of sight the more acute his hearing became. Perhaps this change resulted from straining harder to pick up every piece of sound passing his way, which included a cacophony of wails, moans, and screams, incongruous jolts of laughter, and similar noises associated with a battlefield hospital.

On the morning the duty nurse admitted Heron, she came forward to change his dressings and began peeling the outer gauze from his face then suddenly froze. "Oh my God!" she gasped. Then off she ran to fetch Lt. Leonard B. Bristle.

"This one's really bad, doctor," she whispered.

"They're all really bad," he replied with mild disdain. "Eighteen year-old amputees, paraplegics, stomach wounds so huge we can't stretch the skin far enough to cover them. Tomorrow we'll see hundreds for whom we can do little more than ease their pain."

The doctor's hair looked wild and a stethoscope swung like a pendulum from the pocket of his white jacket as he moved from the opposite end of the tent toward Heron's bed. When he stopped beside a portable sink to remove the bloodstained coat and begin washing his hands, the nurse took note of his eyes flashing fatigue and lines of stress etched in his face.

After slipping on a clean jacket, he grabbed for the stethoscope. "He can't be any worse off than any of the others," the doctor said on the fly. "Last night, or was it this morning, I excised the legs of a young man of eighteen. Today I told another he'd never use his arms or legs again." When he finally reached Heron's bedside Bristle turned abruptly to the nurse and whispered, "Holy shit! This does look bad!"

H E R E L I V E D it over and over, trying to piece things together, groping for a handhold on an icy slope but finding none. The bright flash and the concussive power of the explosion left him unaware as he hurtled through space, shrapnel ripping into his body and white phosphorus burning his flesh, multiple branding irons pressed against his skin, searing to the bone. The experience imbedded itself in his brain like a cancer that would revisit him in his sleep and register again and again in sudden uninvited replays in the weeks and months to follow.

The unrelenting pain from physical injuries left him wondering about damages to his brain; and worst of all how long, he wondered, would total blackness with only occasional flecks of light endure?

"You don't have to do this," Madeiros had argued. It had to be done; the odds simply went against him.

What about Azelia? What will she do? Twenty-four years old and at the peak of her beauty and with a smile that broke down barriers, drew people to her, and warmed total strangers at first glance. Azelia appealed to everyone. Women wanted her as a friend. Men wanted to protect her, and she could have her pick of any man.

Suddenly, he felt a presence. "Who's there?" he asked, his voice so feeble the priest barely made out his words.

"Father Connors, son. How are you feeling?"

"Father Connors? From Fitchburg?" He recognized the steadfast voice.

"The same. Feeling any better?"

"When I regain my eyesight, perhaps." Momentary flashes of light gave rise to hopes of regaining at least some of his sight. His heart lifted at the thought, ushering in a flood of dreams for a brighter future.

"What are you doing here? I mean how did you...?

"What have they been doing for you, Larry? How are they treating your injuries?"

"They administer morphine when I scream and when it wears off give me more. But how did you...where did you come from?"

"A miracle, I guess. I just happened to be in the vicinity on the day it happened. They went looking for a chaplain and found me. They didn't know you can't kill Larry Heron that easily. I'll speak to the doctor to see what else they can do."

Connors soon learned that besides morphine, penicillin and sulfa, the doctors prescribed whole blood, and that Heron remained incoherent the past few days – barely conscious.

Connors located Dr. Bristle and asked, "What's the prognosis?"

"Not good, Father. He's lost so much blood that it's a wonder he's alive. He's still not out of the woods."

"Doesn't look much better than when I saw him on the battlefield, but not much worse either. So what does it mean?"

"I don't want to shatter your hopes, Father, but I'd give him no more than twenty-four hours. Severe burns and blood loss initially sent him into mild shock. But recovery from shock does not alter the long-range prognosis."

Connors fixed Dr. Bristle with tired eyes. Exhausted and out of breath, the priest hadn't slept but a few hours each night for over a week and getting through each day on fumes. "I'd like to pray for him. Give him last rites one more time, if you're sure that's the case."

"Be my guest. Does it matter that he's not awake? I sedated him."

"Even better." After performing the ritual one more time, Connors located the doctor. "What happens next?" he asked.

"Tremendous fluid loss and physiological shock. The phosphorous burned the exterior of his face, especially the right cheek, down to the subcutaneous tissues causing edema."

"These aren't like the usual burns you treat. I mean not caused by fire."

"But similar. White phosphorous burns in the presence of air, and since no one took immediate action to smother the burned areas with mud or water, the damage went deep and took a heavy toll." The doctor sighed. "He has a lot going against him. Plasma leakage, decrease in blood mass, lowered circulation, all contributing to metabolic disturbances. Tissues and red corpuscles releasing liquids into his circulatory system could introduce anemia and hypertension, and lead to cerebral and pulmonary edema."

"English please."

"A build up of fluids and toxic products from destroyed tissues re-absorbed into his blood stream could result in liver and kidney damage. In the next twelve days, if he survives that long, he will drop about ten or twelve pounds from fluid loss alone. But that's not all. There's also the danger of infection which could inhibit nutrition, block healing, and finally result in death."

Fr. Connors approached the bed and stared down at Heron, sleeping restlessly now. He picked up his chart and read: *Severe burns of the cheeks, eyes, nose, a good part of both ears, and lower lip. Burns on forehead, neck, and shoulder. Shrapnel wounds over the entire body.*

Connors planned to offer some very special prayers that night with the hope of shifting the odds in Heron's favor. Throughout the thick of fighting in North Africa, Tunisia, and Sicily, Fr. Connors proved not only a dedicated priest but a caring person. Even before landing on Utah Beach with the 9th Infantry, he administered last rites to so many fine young men that he long ago lost count. Unfortunately, Larry Heron appeared in worse condition than many who did not survive. The chaplain departed from the bedside when a medic came by to change Heron's dressings and tend to his medications.

Connors lay on his cot that night and unable to fall asleep reached for his mug with CONNORS COFFEE SHOP painted on it. Finding the pot nearly empty, he tipped it more than forty-five degrees to eek the last drops from the spigot and watched it ooze out blacker than night but still plenty hot. He took the mug outside his tent where he located a folding chair and sat down to stare up at the stars. Luxuriating in the warmth of the mug, he lifted it to his lips with both hands and sipped the strong blend, remembering back to the 1930s, the way things went down in those precious days and never would again.

ON HIS FIRST NIGHT inside the field hospital, he woke up screaming. It felt as though a fifty-gallon drum had been dropped over his head
with someone pounding on it with a sledgehammer then once again
awakened from the white phosphorous dream, the classic flashback. "A
common experience of survivors following extreme trauma, particularly
burn victims," the doctor explained. "It will go away in time." Amazingly,
in the dreams he always saw things in color, which the doctor claimed
unusual.

He drifted off but awoke an hour later with searing, burning chemicals eating his face and shrapnel tearing through his body; relieved to open
his eyes to cold sweat, he lay staring at darkness. This had better end soon,
he thought. *If only the pain would ease.* Moments later the nurse arrived
with a welcomed shot of morphine.

Fr. Connors dropped by the following morning prepared to pray
for his friend's departed soul. Heron lay in a corner of the tent very much
alive, his face wrapped in heavy bandages as well as his shoulders, chest,
arms, and legs. Perhaps he would pull through after all but what kind of
life lay ahead for this once proud athlete?

Connors came to regard "field" and "transit" hospitals as unnecessary delay points in the evacuation process. Evacuation of patients deemed
"transportable" took place as quickly as possible whereas "non-transport-
able" patients, those like Heron predicted to die, were left unattended until
they either passed away or showed strong signs of recovery.

Connors located Dr. Bristle to ask why Heron no longer received
treatment. "Why aren't you giving him penicillin?"

"Like I told you, we don't think he has any chance of survival, and
even he did, we are under strict orders to give first aid care only in this
hospital."

"What?" The chaplain asked incredulously, the doctor's words taking him totally by surprise.

"Father, there are thousands of wounded men with a much better
chance of survival than Heron and that's where we must focus our limited
resources, using them only on patients for whom they will do the most
good. It leaves us no choice but to keep Heron off penicillin."

"What's the problem with penicillin?" Fr. Connors asked.

"Fifty billion units ordered for June never arrived, forcing us to
ration the 600 million units available during the assault phase of the inva-

sion. Only half those 600 million units actually made it to the beaches while the rest found their way onto LSTs destined for hospitals in England." He paused. "For patients in transit we were given strict orders to apply standard operating procedures only, which translates to administering first aid care and nothing more. Keep those alive we think can make it, load them on a ship and send them to England for proper care. The rest we leave on their own." The doctor's face flushed.

"Leave to die, you mean." The chaplain could not remember ever before feeling this angry.

"You think I like it? Those are my orders, for Christ's sake! Oops – sorry Father."

Connors didn't stay to argue. Instead, he rounded up a jeep and paid a visit to the 9th Infantry Division commander, General Manton S. Eddy. Connors joined the 9th Division in October 1942 and gained Eddy's ear. Connors trained with these men in England just before they landed on Utah Beach on June 10, as their imbedded chaplain. He landed in the front lines right beside the troops when they first saw combat in North Africa on November 8, 1942, and he ran beside them when they launched their attack in southern Tunisia on March 28, 1943. And when the 9th drove to Bizerte on May 7, he shared foxholes with them. In August, he accompanied the Palermo landing to participate in taking Randazzo and Messina, proving a great source of comfort to every soldier in the division.

When Fr. Connors related Heron's story to General Eddy and asked why the wounded were not receiving the care they deserved, the general became furious. He sent word to General Paul R. Hawley, the Army's Chief Surgeon. Hawley blew his stack when he read the message. "That's a gross misinterpretation of my order. SOP does not stand for neglect." Earlier that very morning he responded in similar fashion to a complaint coming from Colonel Cutler.

General Eddy received an immediate response from Hawley: *Orders misinterpreted. New orders cut today. Every effort will be made to save every injured man. Enclosed is (1) a copy of my order and (2) a personal message on its way to Lt. Bristle.*

The next day, Dr. Bristle placed Heron's name on the list of injured slated for evacuation by hospital ship to Salisbury, a ship fully equipped and staffed with highly trained personnel to begin the process of reconstructive surgery.

"Will I ever see again?" he asked.

"I don't know," was all the doctor would volunteer.

On July 2, 1944, Heron spent long hours on a stretcher in a location reminiscent of a scene from *Gone With The Wind* where the wounded lay stretched out side-by-side on the streets of Atlanta for as far as the eye could see.

He reached up and felt his shirt pocket for the rosary beads. "My rosaries!" he cried out. "Where are my rosaries? They were in this pocket."

"We didn't find anything like that," a deep male voice answered. "I'll look around. Is there something special about them? Perhaps we can replace them."

"Those cannot be replaced," he murmured.

The orderly spent a half-hour searching through Heron's clothing failing to locate the string of beads. "I'll get word back to the field hospital and see what they come up with." He feared the orderlies likely removed Heron's shirt when they took him into the tent then discarded it with the beads still in the pocket. Doubtful they would ever turn up. Heron viewed the loss of the rosaries as another bad omen.

Several hours later, orderlies lifted Heron aboard an LCT filled with dozens more casualties on stretchers that transferred them onto a LST hospital ship lying off the coast. The LST would not leave for England until fully loaded, which delayed getting underway another twenty-four hours.

After docking in England, eight additional hours passed before Heron finally reached shore for transport overland to #158 General Hospital in Salisbury. The #158 complex contained 2000 beds distributed throughout small clapboard buildings set in neat rows resembling an Alaskan mining town.

The town of Salisbury, best known for its Salisbury Cathedral, stood as the mother church of the Salisbury Diocese, a glory to God in stone and glass with magnificent 13th century cloisters, slender columns and sculptured busts of dark Purbeck marble set in its sparse, barren interior. Built from 1220 to 1258, this awe-inspiring cathedral provided a powerful yet understated setting for many great occasions over the centuries. The majestic spire rose to 404 feet, making it the tallest such structure in the world when first erected. One of the four original copies of the Magna Carta, remained on display in its 13th century Chapter House along with a stone-carved frieze of Old Testament Bible stories, marking significant chapters in the Cathedral's history.

None of these facts were of consequence to the heavily sedated man transported to Building #28. Upon Heron's arrival, Lt. Vance Bradford arrived to start his examination but paused, horrified to find a patient with third degree burns to the eyes, lids, face, forehead, neck, shoulders, and hands; a young man missing tissue, dead tissue, and festered cells; as well as covered with multiple shrapnel wounds.

#158 General Hospital type, England circa 1944.
Courtesy US Army Medical Department

Heron's condition continued to weaken as his immune system lost ground to infection. The smell of food from a mile away would cause him to retch, depriving his body of protein and a long list of nutritious supplements desperately needed for recovery. "I want him on a liquid high-protein diet, immediately." The doctor continued to write in his chart as he spoke.

"Get me a blood count. Continue the penicillin, 20,000 units every four hours," he ordered. "And start him on sulfadiazine every four hours."

Dr. Bradford reapplied fine mesh gauze soaked in petroleum jelly to the burned areas then added bandages, a treatment developed at Mass General Hospital in the early forties by a doctor whose first name matched Vance Bradford's last, Dr. Bradford Cannon.

As the medications wore off, Heron's senses became more acute and he grew restless, trying not to scream in pain. Dr. Bradford administered a shot of morphine and prescribed additional doses four times a day. "Let's get some head X-rays," he told the nurse, while jotting more findings on Heron's chart.

After re-examining the shrapnel wounds, Dr. Bradford ordered saline sulfa dressings bound to his right shoulder. That evening, in addition to morphine and changed dressings, Heron received a 500 cc blood transfusion.

THEY STOOD together in a large room gazing through the slits of a venetian blind and out a window overlooking a river of dark blood. Azelia pointed to a dark silhouette approaching them and gasped, knowing just as he did, that the evil being wanted them dead. He slammed closed the blinds to shut out the Devil, and as he did a hand closed over his mouth and he began to suffocate.

Suddenly, he tried to rise to a sitting position and pry open his eyes but could not dispel the darkness. Vertigo seized him as hands grasped his arms and a gentle voice said, "It's all right. One of the pillows covered your face and stifled your breathing. Anyway, it's time for some breakfast, if you feel up to it."

Heron moaned at the thought of food. Instead he asked for and received a shot of morphine followed by sulfadiazine then the doctor arrived and ordered another blood workup.

As the drugs began to take effect, Heron asked the orderly, "What day is this?"

"The 4th of July," the orderly answered.

Heron wondered if his comrades still engaged in the fighting were at that moment faring any better.

SEVERAL YEARS passed before the woman in Milford Hospital's Room 201 would discover what became of David Rubenstein and a few of the others whose names she learned over the years.

A tray crashed to the floor somewhere down the hall, the sound jolting her from her reverie. Moments later she heard a door close and silence prevailed upon her once again. She looked straight ahead to find the moon had progressed slightly on its path without her, but still plenty of time left.

Now where was I? Oh yes, David Rubenstein and Heron's other comrades, and rightfully so.

On that particular 4th of July men of the 87th Chemical Weapons Battalion suffered some of their worst tragedies.

COMPANY A began July 4th firing in support of the 2nd Battalion of the 330th Infantry Regiment. Not long after Heron performed his heroic deed, Cherbourg fell to the Americans but at a steep price. Six officers and 88 enlisted men of the 87th died, and following the assault, Company A continued on its trek non-stop through towns that read like a French tour guide.

Thus far in the campaign, the company fired off 32,000 mortar rounds, and following a lengthy barrage on the town of Greville, 150 Germans came forward to surrender. The push to Carentan, found their forward observers operating side-by-side with counterparts in the field artillery, combining efforts so that their forward observers could supply fire direction to not only their own mortar units but artillery batteries as well.

Following a 200-round volley in the early morning hours of July 4th, there came a lull in the fighting that provided time for a well-earned breakfast. Most gravitated to their respective foxholes dug under tall trees for protection where they broke out their mess gear.

Roger Burt sat sipping coffee when Francis Healy leapt to his feet and challenged him to a wrestling match. Healy danced around jabbing the air like a prizefighter. "Come on. Let's go a few rounds." The two men engaged in wrestling between combat missions to help ease tension and stay in shape.

"Are you nuts?" Burt asked. "I haven't finished my coffee."

"Afraid?"

"Of you? You're crazy." Rising to his feet, he added, "Let's go. But remember, no gouging, kicking, biting, stomping, or pulling of hair."

"What do you think I am, an animal?"

"A sissy."

"You're begging for it."

They moved off to a clearing about fifty feet from the cluster of foxholes, circling one another like angry cats, eager to slough off pent-up energy. "Gotcha," Healy boasted triumphantly, as he clamped his hands around the back of Burt's neck. But the wiry Burt grabbed a handful of shirt and fell backward pulling Healy with him and using leverage to flip them both with him landing on top straddling Healy and pinning him to the ground. He began counting, "One, two, three..."

Kaboom! An enormous explosion shattered the quiet as an enemy shell exploded in a tree directly above where the men sat in their foxholes. Some sat beneath trees that offered no protection, and in fact contributed to the downpour of shrapnel raining down on them. *Kaboom! Kaboom!* Instead of wearing their helmets to protect their heads, most had left them sitting turtle-like on the ground or inverted in their laps to hold rations or water for washing.

"Tree bursts," Healy shouted. The two wrestlers covered their ears and hugged the ground as several more shells exploded in nearby trees. When the shelling ended, a few seconds of ominous quiet preceded several screams and groans, proof that the German forward observers were scoring deadly hits. Rising to their feet, Healy and Burt ran to see what they could do for the injured.

They came first upon Pfc. Sheffield lay on the ground, crying, "I've been hit. Oh God. I'm bleeding."

"Jesus!" Burt exclaimed. Sheffield's arm bled all right, but when he wiped away the blood all he could see was a tiny scratch. He turned away and sank to his knees beside Staff Sergeant David Thomas. "Oh no. Not you," he said. One look told him he could do nothing to help his close friend who sat Buddha-like, mess gear in his lap, a spoon in his hand, a single pinprick of blood oozing from the side of his head where shrapnel had entered, the only wound Burt found on his body. Burt took him gently by the shoulders and lowered him to the ground then reached over to close his eyelids and in the process, his tear-filled eyes went straight ahead to the spot where Thomas' dark lifeless eyes had been staring moments ago. There first time he noted the bodies of Sgt. Volcjak and 1st Lt. Arthur L. Gump lying in one tangled heap, like a pair of discarded puppets. Without further examination he knew both were dead.

Cpl. David Rubenstein sat slumped over in his foxhole, the top of his head sheared off while in the act of opening a can of rations. Pvt. Leslie S. Kolman's crumpled body lay beside him, a pool of blood outlining his head.

Sheffield continued to sob, "Help me! Oh God, help me!"

Burt decided he'd better take a closer look at Sheffield's wound. "It's just a scratch," he said, dispassionately. "A million dollar scratch."

Despite his words, Sheffield believed he suffered a life-threatening wound and continued to moan and wail.

Burt and Healy split up and went to search for help. Burt soon

located a handful of medics sitting beside two chaplains, all eating breakfast in a barn a hundred yards away. "Come quickly," he said. "They need
help."

Heads turned and reading the terror in his eyes, each rose to his
feet one by one and followed him wordlessly. Five dead, seven wounded,
including Art Fertitta and William S. Collins, who would later succumb to
his wounds. Interred in France, he would remain buried there until 1947,
when finally his body came home.

"How's Sheffield doing," someone asked Healy.

"Shove a lump of coal up his butt; he'll squeeze you out a dia-
mond. That's how tight he's wound but absolutely nothing's wrong and
he'll be fine."

The next afternoon fared no better for Company A. The 3rd Battalion of the 329th Infantry Regiment led an attack south of Piereres at 0700,
with Company A getting off 300 rounds of supporting fire before moving
to a position south of Le Culet.

Lt. Branson and Lt. Moore went out as forward observers to relieve Lieutenants DeWitt and Berry. Lt. Branson, the third officer selected
for forward observer duty, located Fiske. "Cpl. Fiske, grab your Browning
and follow me. You too Shanahan, we need a good radioman." Shanahan
smiled and eagerly bent to the task of adjusting the gain and squelch dials
then hefted the radio pack onto his back and followed.

The three men moved forward about a thousand yards to the base
of a hill that appeared a perfect spot for an observation post from which to
direct fire on enemy positions then report back on results.

A German machinegun team, concealed behind a copse, spotted
the men as they started up the hill. "Was uns hier haben?" said a German
soldier peering through binoculars at the approaching men.

The gunner, looking through his own binoculars, responded,
"Amerikaner!"

The Americans continued forward, spaced about ten yards apart.

"Kommen Sie hier!" The gunner whispered, coaxing them closer.

"Hold up. Something's not right," the lieutenant said. "Too quiet."

At his signal the three men dropped to their knees to listen.

"Call for fire on that hill," Branson said.

"Roger." Shanahan rose to his feet to slip the radio off his back.
That's when the machine gunner raked a path across his chest that
coughed holes in the radio. His hands flew out to his sides like a chicken

flapping its wings. His body danced grotesquely. Mini-geysers of red sprang from his chest and soaked into his clothing. If he screamed, no one heard above the clatter of the machinegun as his body smacked the ground like a lead-stuffed doll.

"Shit!" cried the lieutenant, diving for cover behind a small rise. Fiske made a running dive into a crater, landing to the lieutenant's left, as bullets kicked up sand at Fiske's heels.

The German gunner focused on the mound that afforded Branson only minimal cover, and when his machinegun coughed the lieutenant felt an immediate stab of pain in his upper left leg and at the same time a burn in his left shoulder, and knew he'd been hit.

A smile creased the gunner's face as he continued to squeeze off rounds. Bullets kicked at the mound protecting the lieutenant, eating it away, slowly chewing it to dust. Branson lay powerless to do more than pray and count down what appeared to him, the remaining seconds of his life.

Left alone and temporarily ignored, Fiske rolled to a kneeling position to lay down a withering fire with the Browning, catching the enemy off guard. An expert rifleman, he had hunted often with his dad. Before the Germans could respond, he regained his feet, advancing and laying down rapid fire, the rifle pinned against his side by a powerful forearm.

John M. Browning designed the weapon for trench warfare and that's exactly how Fiske employed it, firing with such deadly accuracy and rapidity that the Germans could only duck for cover. One bullet struck the German gunner in the forehead sending brain matter spewing out the back of his skull. The second frantically pushed his comrade aside and took up the Mauser but it jammed. Panicking, the second gunner lifted his rifle and fired blindly.

A bullet tore a path near Fiske's heart and another gushed blood from his left shoulder. As he fell to his knees, Fiske removed a grenade dangling from his chest, yanked the pin and lobbed it toward the machine-gun emplacement. As the grenade sailed through the air, another bullet caught Fiske in the throat. The last thing he heard as he pitched forward to meet the ground was the sound of his grenade exploding.

Lt. Branson, only slightly wounded, ran to Fiske's side firing his pistol in the direction of the Germans. He picked up the Browning but all activity in the machinegun nest had ceased because both Germans lay dead.

Branson felt sick to his stomach when he couldn't get a pulse. Hoisting Fiske over his shoulder, he limped down the hill and laid the body behind a clump of trees, then returned for Shanahan. As Shanahan's limp body rolled next to Fiske's, Branson sensed activity and glanced up the hill in time to see a number of Germans advancing toward his position. Nothing left for him to do now but get the hell away and live to fight another day.

THE FAMILIAR rumble and rattle of the blood cart reaches her ears as the door swings open and a black man she nicknamed "the vampire'"enters her room. "Good morning, Mrs. Heron."

"You're late. Usually you get here at 1:30 a.m. and it's now a quarter to 3."

"Yeah, I know," he says patiently. "Had a little car trouble this morning but everything's fine now. I'll make it quick for you."

"Oh, that's all right. I wasn't sleeping anyway."

"Now let's take a look at you." Dropping a pile of blood-letting paraphernalia on the bed, he takes her arm in one of his hands and taps on it a few times then wraps a rubber piece around her upper arm. "Make a fist."

Familiar with the drill, Azelia balled up her fist when she saw him coming. She hardly feels the needle going in except for a tiny prick then before she knows it, he wheels the cart toward the door.

"Have a good night, Mrs. Heron."

She finds herself alone with the door closing behind him before she can answer.

The clock beside her bed reads 3 a.m., and the moon now shines in at her from beyond the half-way point in its trek across the window. As she stares ahead, a strange feeling passes over her and she can feel her body growing weaker, like everything inside her wants to shut down. It's her body telling her to let go but her mind flat out refuses. Not now. She must not doze for if sleep overtakes her, she may never wake up again.

Where did she leave off? Oh, yes.

EACH TIME Heron learned something new about the ill fate of a comrade, it took many days for him to get over it. Though she never met many of them, she knew them all by purposely gathering details about each of them over the years, about the ones that made it home and those that did not. And death did not always choose to visit just the members of the lower ranks.

On July 12, 1944, while the fighting and dying continued on the battlefield, Brigadier General Teddy Roosevelt, Jr. dropped dead of a heart attack while serving as the military governor of Cherbourg. He left behind his wife, Eleanor, nicknamed Bunny, and their four children. Azelia read how the couple built their marriage on a firm foundation of trust and deep love, not unlike her relationship with Larry.

Not long before he died, Teddy sent his wife a letter summing up their "grand life" together, how they packed enough into the years "for ten ordinary lives." Before he died at age of 56, Teddy Roosevelt, Jr. wrote of their happiness. "I pray we may be together again," he wrote.

The dying didn't stop there.

On July 13, Lt. Bonafin called down to the battalion commander of the infantry unit for which he had been directing fire on enemy positions in the vicinity of Bois Gommet. "It looks to me like a rattrap, sir," he warned. "They're making it look too easy." The commander disagreed. "We have them on the run, is all."

Sometime later, Bonafin heard the dreaded squeaks and clanks of tank treads. Mark VI tanks suddenly burst into view one after another from behind camouflaged positions to encircle the American infantry battalion. Caught in a rattrap along with the infantry unit he tried to warn, the lieutenant lost his life to a hail of bullets from a machine gun mounted on one of the tanks. He died instantly.

"They were all so young," she whispers to herself. "Like my Larry."

CHAPTER THIRTEEN

NO GOING BACK

EACH MORNING, Heron took a light breakfast and almost every day before someone wheeled him to the operating room for a rough scraping and bathing of burns, a process called debridement or enucleation, an extremely painful method of removing dead tissue by brushing or peeling. Treatment and dressing of his shrapnel wounds then followed.

On July 14, Dr. Vance Bradford visited his bedside with a heavy heart, his voice gentle. What he came to do, he had never signed up for. He did it to save lives, help patients heal, cheer them, not bring them terrible news. "Sergeant," he whispered.

"Yes."

"It's time we discussed your eyes."

"Did the nurse tell you? I think my eyesight is returning!" Heron sounded elated.

The doctor took in a deep breath. "You told her that you see flashes."

"Yes. Doesn't that mean I'll see again?"

"I'm afraid it's not possible."

"What?" Heron's voice grew feeble, weighed down by shock and disbelief.

"You see," the doctor said, quickly regretting his poor choice of words. "Your eyes were completely destroyed, burned away by the intense heat. Phosphorus particles clinging to his eyes and body, burned through the flesh and left deep, ugly craters. "There's no way of saving what's left of them."

"But the flashes?"

"A sporadic residual retinal reaction. Electrical impulses from the optic nerve. Nothing more."

It could not be true. Any minute now, he would awaken from this nightmare. It could not really be happening. Desperation seeped into his voice. "Can't you fix them?"

"Your eyes represent dead tissue that must be removed before infection sets in."

Heron's sigh reflected an overwhelming hopelessness. "When?" he asked, as if it made any difference.

"The sooner the better." A pause. "Tomorrow morning."

This was it then. Nothing would ever efface the darkness. Hope vanished. Until this moment he held onto the belief that his sight would one day return. Whenever a doubt crept in, he refused to accept it. But reality now set in.

When left alone, he wept, but no tears fell from his dead eyes.

AT 0730, NURSES wheeled Heron back to the operating table to prep his back, buttocks, thighs, and abdomen to use as donor sites. At 0900 he underwent eight-hours of surgery to remove what remained of both eye-balls. Dr. Bradford covered the empty eye sockets with split skin grafts, and covered as well parts of his forehead, face, ears, and neck.

The doctor checked in later that night to find his patient lying on his right side, his left, the donor site, covered in bandages. Heron woke up complaining of severe eye pain. The doctor increased his morphine.

"Keep him on a liquid diet – whatever he can tolerate," he ordered. Soon after the doctor left, two units of plasma and 500 cc of blood were administered, and his blood pressure taken every fifteen minutes for the next three hours.

The next day, the doctor found Heron awake and quite lucid. "You're a very brave man," he told his patient. "I grafted the soft defects in your right cheek and lower lip. The wounds will soon heal but you'll need many more operations before things begin to look normal. That will take place back in the states under the hands of our best plastic surgeons."

"In a few weeks, when the sites have healed, we'll send you on your way. Meanwhile, I'll introduce you to some Red Cross workers who will teach you to read Braille. Would you like that?"

Heron nodded his approval.

"Oh, by the way," the doctor added, "This arrived for you today."
He put something in Heron's hand.

"My rosary beads? They found my rosary beads?"

"Appears so."

"Will you describe them to me doctor?"

"Sure. Black beads with a solid silver crucifix."

Heron tried to smile but his face felt tight, like skin stretched over
a snare drum.

"There's a note that came with them." Dr. Bradford read it to him.
*I am the orderly who searched for your rosary beads and finally tracked
them down in a pile of clothing headed for the incinerators. Hope they
bring you luck.*

"I don't even know his name so I can thank him."

Bradford said, "It's signed by a Ronald Williams. You can dictate a
thank-you note so the nurse can forward it to him."

"I can't believe he found them."

How awful it must have felt for him lying in #158 General Hospital,
though he never complained to her about it. She learned later that he spent
his time practicing Braille and reminiscing about his "first life," a life that
included her in it every day. She returned to those days in her mind, the
Friday night dates that invariably ended with a movie at the State Theater
in Milford. Following the movie, they would cross the street to Nolan's
drugstore. She could almost smell the apothecary odors emanating from
the drugs on the shelves. The drug store included a soda fountain with a
granite counter and marble-topped tables surrounded by hourglass chairs.
They'd sit at one of the tables and eat hamburgers or fried chicken then
follow it down with a slice of homemade apple pie or banana split.

On Saturdays they might go to the park and play tennis in the
afternoon, on the perpetually well-maintained clay courts. Rainy Sundays
would find them at a friend's house or at either of their homes playing
Monopoly. And in the dead of winter, when the pond froze over, town
plows arrived to clear away the snow for ice skating and "pick up" hockey
games.

Life in Hopedale seemed simple then, a time when a dollar bill
would buy enough meat to feed a small family. Just one dollar would

buy four large potatoes, a pound of green beans, a head of lettuce, a large tomato, and four dinner rolls, with five cents left for a Coke or Pepsi. They led safe lives before all this and enjoyed a comfortable existence.

The awful war took him to foreign shores and just like that, in the blink of an eye he became the war's most severely injured veteran. She cried now as she pictured him lying alone in his bed surrounded by what had to feel like a tomb of darkness, knowing he could never pick up life where it left off before the war, knowing he could never again play the sports he loved. How frightened and alone he must have felt, though the doctor's report she obtained years later indicated that he handled it all like a warrior.

PURSUANT TO ORDERS that he be "remanded to the Zone of the Interior" (Continental U.S.) on August 15, 1944, Lawrence J. Heron was taken aboard a ship that transported him to New York. Before his departure, Dr. Vance Bradford wrote the following summary into his records: *At 1830 hours on 26 June, 1944, at the battle for Cherbourg, France, Heron was unloading phosphorous bombs and was severely injured when enemy fire hit and exploded a bomb he was carrying.*

The day after admission to #158 General Hospital in England, on 3 July 1944, I had him in surgery without anesthesia to deride necrotic burn tissue from his face and neck. His eyelids were practically destroyed and his eyeballs were without color, resembling rotten grapes. It was my unpleasant duty to tell him that his vision was gone. He told me that he could see some flashes, but of course what he thought was some vision was only sporadic residual retinal or optic nerve reactions.

At the 8-hour operation on 15 July 1944, what remained of his eyeballs was removed and the whole forehead, face, ears, and neck were covered with split skin grafts. One strip of skin graft covered the empty eye sockets.

The grafts healed well without complications. Although only 24 years of age, Heron accepted his blindness with exceptional serenity.

While waiting for transportation to the Zone of the Interior, Heron was already being rehabilitated. With a cane and a guide, he walked about the hospital grounds, visited Col. Graham, our commanding officer, and was coached in Braille by the Red Cross women.

Heron has soft defects in his right cheek and lower lip. The wounds were grafted and healed but he will need more plastic surgery to restore

*these defects. Although regulations permitted us to do any cosmetic sur-
gery that could be accomplished within 180 days, our basic mission was to
get the wounds healed and preserve function.*

*Heron is quite a stark contrast to another burn victim with severe
disfigurement of his face and eyelids. The other man repeatedly tried to
commit suicide, even though he had not lost his vision.*

The war affected the lives of so many people in countless ways. It
would never be the same for any of them after the war. Roger Burt long
since told her stories of how the war altered his life, especially how it
strengthened his great faith and belief in God.

CHAPTER FOURTEEN

TRUE TEST OF LOVE

BURT STARED up as wave after wave of fighter-bombers passed overhead dropping loads as close as a quarter mile ahead, so close that it raised the hackles on the back of his neck. The bombing continued all day, cutting a wide path for the infantry to move through. This would go down in the history books as the operation preceding the Saint Lo breakthrough, one of the most bloody and difficult phases of the Normandy campaign.

As he entered the small village ahead, a child no more than eight years old came up a flight of stairs from a basement apartment and moved toward him holding something small in her hand. As she drew closer, she held out a small gold cross obviously wanting him to take it, so he reached out and lifted it from her. "Merci," he said. Then he thought back to a simple phrase from the pages of his French language book and asked, "Quel est votre nom?"

"Michelle," she responded.

"Mon nom est Roger," he said. "Roger Burt." He squatted down to bring their faces to the same level.

She smiled and kissed him on the cheek then without uttering another word turned and walked briskly down the block to return to her apartment.

He held the cross for a long while, turning it to catch the light, then, regarding the token as a sign of protection, he slid it into a compartment of his wallet in front of a booklet of prayers given to him by the nuns of St. Joseph, prayers that whenever read would continue to protect him from harm for at least the next thirty days. With these two precious items in his wallet, he felt safe, and would keep them with him wherever he traveled for the remainder of his life.

ON THE MORNING of September 12, Company A, firing from Koul in support of the 1st Battalion, 16th Infantry of the 1st Division, dropped twenty-five rounds on an enemy convoy causing the Germans to flee in panic for the nearby hills. Moving through the Aachen forest, they came upon the first of a series of fortifications that guarded one of the main approaches to Germany, a series of hills and ridges marked by marble blocks, wire fences, and tall trees. The men continued their push along a road that followed the valley floor built to connect local towns and villages to the industrial mining centers of Forbach, Stiring-Wendel, and Saarbruecken. This region was known as the southern gateway to Germany's Sarre Basin, the fortress city of the Siegfried Line. Though the soldiers arrived in a state of exhaustion, they summoned the energy to dig in. Meeting only light resistance, they took few precautions, unaware that a terrible fate awaited them.

Two hours past the stroke of midnight on Wednesday, the roar of German 88's and German flak guns broke the calm, catching Company A completely by surprise while they lay sleeping. The horrific shelling continued non-stop throughout the night and did not let up until ten o'clock the next morning. During this violent fire storm, a dozen successive shells hit Company A's gun emplacements, causing heavy casualties.

The wounded laid all about as shells continuously whooshed overhead to explode all around Pfc. Angelo Bastoni who ran zigzagging for his life through the fires of hell to reach 1st Lieutenant Doug Peterson, who lay bleeding profusely from a shoulder wound. Bastoni covered the wound with a compress, doing what he could to calm Peterson then went to the aid of other fallen comrades needing assistance.

Ten yards away Sgt. Chuck Learned lay writhing on the ground and six yards further along, Pvt. Almeida lay losing copious amounts of blood through several wounds. Knowing he must act quickly to transport as many wounded as possible before the shells found them or they bled to death, Bastoni dashed across an open field with just one thought, how to evacuate these badly wounded men to safety. He found his answer parked under a tree about a hundred yards away - Peterson's jeep.

Five-foot-eight Pvt. Bob Williamson lay recalling General Roosevelt's words during the landing. "Why are you men wasting your energy digging foxholes?" he asked. "If one's going to get you, it's going to get you. Just keep moving!" And then it happened. An eighty-eight shell exploding ten yards away lifted him off the ground as shell fragments knifed through his body.

Roger Burt thought, *I'm going to die*, as German artillery rained down around him and looking up, saw young Williamson fall with blood pouring from wounds like wine from a barrel shot full of holes. As he made his way toward his fallen comrade, a jeep pulled up beside Burt with Bastoni at the wheel. Bastoni had already safely evacuated Peterson, Learned, and Almeida, and came back for Burt and Williamson. "Come on, get in!"

Burt leapt into the passenger seat and the jeep lurched twenty yards forward to pull up alongside Williamson. Without a word, both men proceeded to drag the man onto a raincoat then lifted his limp body into the back of the jeep, thankful he remained unconscious and felt no pain. Despite Burt's frantic efforts to plug the wounds, by the time they reached the aid station, a gallon of blood had pooled in the folds of the raincoat and Williamson lay dead.

Bastoni had made four round trips before stopping to pick up Burt and Williamson. He drove the jeep through a raging hell storm to remove a total of ten wounded men with Pvt. Williamson the only man lost. For his heroic deeds that day, Bastoni received the Bronze Star. Roger Burt's bronze star and promotion to Staff Sergeant would come in the weeks ahead for numerous acts of courage under fire.

Williamson's death that day came as no surprise to Burt who the night before felt an inexplicable sense of his young friend's impending doom, and it frightened him more because he had experienced similar premonitions prior to the deaths of two other comrades.

Often soldiers witnessing death on a daily basis become more religious with time, and in Burt's case more superstitious as well. Some superstitions made perfect sense like never three on one match because a sniper can zero in on the light by the time it reaches the third cigarette. But no one could explain the frequency of soldiers dying soon after lifting a memento from a dead German. Because it had proven true on so many occasions, Burt, like many other comrades chose to pass on lifting a coveted Luger from a dead German, walking on by without giving it a second glance.

The battlefield proved a robust environment for emotions to run high and a place where men of reason and intellect found religion, accounting in part for the long prayer-lines formed immediately prior to each battle. Whenever a chaplain appeared, it didn't matter which faith, large numbers of soldiers gathered around to pray, and the man drawing the largest crowds was Fr. Edward T. Connors, with soldiers and local townspeople flocking to his Masses.

Fr. Connors sent the following letter to Bishop O'Leary on September 14, 1944: *The people are happy to see us. Occasionally, I hold Mass in a church for my units. When I do, the whole town turns out, though the only warning is the ringing of the bell before Mass. The Nazi's have spread false propaganda that the American army is filled with Communists. But that perception is changing and we are drawing larger crowds at Mass and Holy Communion.*

In one town we entered, the people were in a turmoil of bitterness and grief. The Germans had departed a few hours earlier - after setting fire to many homes. When some men tried to extinguish the flames, they were shot down then butchered. I counted fourteen bodies – a horrible sight. The next morning I offered Mass in their parish church for the victims and those of my men stationed nearby attended, as did the entire town. The SS troops had murdered fourteen men. American troops prayed for the men.

AZELIA HEARS footsteps approaching down the hall and hopes it isn't the nurse heading for her room to disrupt her train of thought. The footsteps grow louder then move past her door and on down the hall where they stop before she hears the elevator doors open and close. Good, she can get on with her thoughts with everything still fresh in her mind.

The men returning from the war all had stories to tell her, but none came back with injuries close to matching what Heron suffered. Newspapers even listed him as "World War II's most severely wounded soldier," as she would learn for herself on that dreary day in August. Everything looked so hopeless for him at the time, and to think she never saw it coming.

THE EIGHTEEN MILE drive to Framingham only took a half hour. She would walk everywhere around town to conserve gas, and both her sisters as well as friends contributed ration stamps so she could make the trip to the hospital each evening to arrive at the start of visiting hours.

As she opened the door to enter his room one night in September, he felt the breeze and smelled the first cool fall air gusting through an open window, followed by the scent of her fragrance just before she took his hand and planted a kiss on it. She seemed anxious, so he let her speak.

"I have something for you, sweetheart."

"What is it?"

Instead of providing an answer, she plugged in the combination radio/record player and dialed a local music station. "I thought you might enjoy listening to some of your favorite programs." She took a seat beside the bed. "Also, I picked up a few records, since I know you love music."

"Thanks, sweetheart." Heron needed something to take his mind off blindness and chronic pain. Perhaps this would help.

The next morning, after the nurse changed his dressings, she asked about the level of pain. "On a scale of one to ten, how would you rate it?" she asked.

"I feel fine," he lied. He hurt all over, worse than ever and would rate it a twenty at least. He requested the doctors cut back on his morphine and now that they did, wanted to scream. After the nurse left, he turned the music louder and instead of screaming began singing along with a Frank Sinatra recording of *I'll Be Seeing You.*

The louder he sang, the better he felt, so he continued singing and felt even better. Every time music played he sang. When he couldn't find it on the radio, he would lay down a platter and sing along. Soon he spent more time in a day singing than sleeping.

Remarkably, singing went a long way toward easing pain, so instead of letting out a scream when it hurt so badly, he would break out in song. *I'll Be Seeing You* rose to sit atop the "Hit Parade" and played every few hours on the radio, and soon Heron knew all the words to it and many other favorite songs. In the past, he felt he lacked time for singing, and really felt no inclination to do so. But now, with time on his hands, he found it lifted his spirits and exercised his lungs. His doctor began encouraging him to sing, claiming it aided healing.

On her next visit, Azelia brought a stack of record albums and in no time he learned the words to *Danny Boy, Irish Lullaby, and I'll Take You Home Again Kathleen.* One Friday night, at the conclusion of *Danny Boy,* applause and cheers broke out amongst the several nurses and orderlies gathered outside his door to listen.

"You have a gorgeous voice," A nurse told him. "How about *Black is the Color?* I heard you singing it the other day and loved it."

Why not? He began: *But Black is the color of my true love's hair...* His voice reverberated along the corridor, and walking patients as well as members of the hospital staff stopped outside his door to listen. Not seeing people's faces prevented an attack of nerves so he really let it rip, and when he finished dead silence greeted him, and he promptly mistook that people hated his voice.

Oh no, *No one likes my singing.* He felt his face flush with embarrassment. If he could see, he would not find a dry eye within hearing range. First the nurse began clapping then gradually everyone joined her, and soon the hall resounded with applause and cheers. From that day forward requests followed one after another, and with each new song he grew bolder and his voice stronger.

As word of his talent spread, groups of patients and staff that gathered outside his door grew larger until he received an invitation to come to the rec-room every afternoon to entertain. His strength grew with each passing day and before his stay at Framingham came to an end, Heron had developed an amazing repertoire of songs committed to memory, and with his newfound talent had achieved an important first step toward recovery.

WHEN AZELIA arrived at the hospital for her visit on the last day before his transfer to a better equipped hospital to receive care from one of the nation's foremost plastic surgeons, he reached out for her hand. She took it in both hers then held it to her lips. "This whole thing must be an enormous shock to you," he said. "I'm sorry for coming home to you in such a mess."

"Sorry? You have nothing to apologize for." She longed to help him understand how her heart ached for him.

"I know I'll never look like I once did, that I must live with blindness for the rest of my life," he said. "But you don't have to. When these bandages come off, you will feel you married a monster."

"Not true. You'll look just fine."

"No. Wait. I've been doing a lot of thinking and we have to talk about this." His voice took on an unusual pleading tone, as gentle and calming as a violin concerto, which stopped her protest before it began.

"You are still young and could have any man you want. If you decide to leave me, I...will understand."

Azelia felt such a shock, she could not speak.

"Before going blind, I would have had to be blind not to have noticed how other men look at you." So beautiful and only twenty-four years old, still young enough to find a new life. If she decided to stay with him, it must be without pity. To get through the awful months ahead, he must erase all doubt. Once the operations start, he could not survive losing her.

"Leave you? Why on earth...?"

"The doctors warned that reconstructive surgery...well, I don't want to travel that road if there's even the slightest chance that when the going gets rough...By the time it's over and you see what I'll look like, well you'll..."

She finished his sentence for him. "Leave you?"

"I'm not the man you married, Sweetheart. No one would blame you. This is not what you bargained for."

"Neither did you. And I don't care what other people think. You're the only man I've ever wanted and I love you more now than ever. So let's get you home where you belong and pick up where we left off!" The tears flowed freely now.

He drew a deep breath. "Nothing can ever be the same. When you see my face, you'll be repulsed. You won't want to hold me close or kiss me, let alone make love to me." He hesitated. "Don't make this any harder than it already is."

Azelia's heart ached like never before. "A Greek poet once wrote, 'Love built on beauty, soon as beauty – dies.' Our love, built on a strong foundation will endure."

"There's a rough road ahead. The pressures will prove enormous."

"For better or worse. In sickness and in health." She stabbed at her eyes with a tissue.

"Sweetheart, an ungodly number of painful operations lie ahead for me, just to make me presentable, not to restore my looks, not even to make me look normal. For you to change your mind and quit part way through would hit me harder than if you just walked away right now." *I can't face years of operations without you, and without you nothing matters.*

She played the words in her head before speaking, like swishing a fine wine around in her mouth to sense the flavor and aroma before swallowing. "If you were to decide against the operations, if you want to stay looking just as you do right now, it won't matter. It was never just your looks that attracted me. It's what's inside. Can't you see that? I want to grow old with you, Larry Heron. That's all I've ever wanted. I've loved you since I was six years old, wanted your children. Still do."

She pulled another tissue from beside his bed. "You're stuck with me and that's that!" She leaned forward and rested her hands lightly on his shoulders wishing she could embrace him.

"I love you, dearest," he whispered. "Always have, always will. It was you who kept me alive, the thought of you, wanting to be with you. But now..." He hesitated. "Now I just want to spare you a life of misery."

"Oh Larry. I love you so much. I would die without you." Impossible now to stop the tears from flowing, she wiped at her eyes again and pressed herself against him, careful not to hurt him.

"You make me so happy. Because of you, I am the luckiest man alive," Heron said tenderly.

A knock and the door opened.

"Just a little longer," Azelia pleaded before the nurse could ask her to leave. "Please?"

The nurse knew she just interrupted something important. "Certainly," she said, "another few minutes won't hurt." Then she promptly left them alone.

CHAPTER FIFTEEN

THE DOCTORS AT VFGH

ON MONDAY, November 6, Heron arrived at Valley Forge General Hospital in Phoenixville, Pennsylvania, three hundred miles further from Azelia, at a hospital an order of magnitude larger than the comparatively small Cushing Hospital, where by now he not only knew his way around but also the first names of most of the staff as well as those of many patients.

Ten days later, he lay alone in his bed immersed in a deep depression when he heard the nurse announce, "There's someone here to see you, Sgt. Heron."

A familiar voice said, "Hello, sweetheart."

"Azelia?"

"I decided to surprise you and booked a hotel for the weekend."

Heron felt his heart rate skip a beat. "Why didn't you tell me you were coming?"

"And give you the chance to tell me not to come?"

Didn't she know he would never discourage her from paying him a visit?

She took his hand and leaned in close. "Can you sit up?" she asked.

"Yes. I can walk around and do anything but see."

"Anything, huh? No, I mean will you sit up for me right now?"

"Sure. What's up?"

Without answering, she helped him to a sitting position and started untying his hospital gown.

"The nurses," he protested.

"They won't come in unless you push the alarm button. They promised me an hour, so we better get started. She carefully helped him undress then slipped under the covers. He felt her naked body press against his and let out an audible sigh. "This beats the hell out of submersion in a saline bath," he exclaimed.

"Let's not waste time talking," she whispered.

ON WEDNESDAY November 8, 1944, a blustery cold wind from the north deposited cold wet rain on Valley Forge General Hospital, while inside First Lieutenant Joseph E. Murray removed his Harvard Medical School ring and the watch his wife presented him when he graduated from Harvard Medical, placed the items in his suit jacket then draped the jacket over a hangar and hung it in his locker. Col. Bradford Cannon entered the room offering a cordial, "Good morning," and took a final sip of coffee before placing his mug on a table beside Murray's.

"I hope we make it a good day for this patient." Dr. Cannon slipped off his jacket and tie then dropped his ring and watch into an inner jacket pocket. "Yes, another sad case," he continued. "The patient's burned over so much of his body that we've run out of healthy tissue and forced to use cadaver skin as a temporary surface."

Scrubs donned, loose hair shoved under caps, masks slipped on to cover up breathing passages, knee controls pushed to turn on water and foot controls to pump drops of antimicrobial solution onto their palms, and then the sound of vigorous brushing broke the silence as the doctors began swiping at their nails.

"If only we could find a way around rejection," Murray said almost to himself as he worked up a foamy lather on his arms. "Think what we could do for this patient." Murray could not shake the habit of devoting large chunks of time probing the mysteries of science, delving into patient's charts and files, running lab experiments, and relentlessly dogging for answers to questions that had baffled the world's best doctors and scientists for generations.

"That's all I think about," Cannon responded. "I wish we could figure a way around it."

The slow rejection of foreign bodies weighed heavily on Murray. "How does the host distinguish another person's skin from his own?" he wanted to know.

After a thorough lathering and vigorous brushing, the doctors rinsed, taking care to keep their elbows low.

"Let's not forget Brown's success with the twins," Cannon said. In 1937, Dr. James Barrett Brown, chief of plastic surgery at Valley Forge, experimentally cross-skin grafted a pair of identical twins.

"Yes. He established that the closer the genetic relationship between donor and recipient, the slower the dissolution of the graft."

Gowned and gloved scrub nurses held out towels for the doctors who followed the prescribed procedure of employing one side of the towel and then the other to dry each arm.

Dr. Murray pushed open the OR doors with his backside and felt the hissing air escaping from the pressurized operating room. He continued holding it open for Dr. Cannon to slip through.

"That was an important first step, a transplant between two people immunologically close – and it worked." That thought would occupy Murray's mind, forming the impetus of future study and dramatic breakthroughs.

"I'm afraid new discoveries and innovations must come from brilliant minds like yours," said Cannon. "It's all I can do to keep up with the flood of patients waiting to get into OR."

"I'm flattered, but I also don't believe it for a minute. You're always breaking new ground."

The scrub nurses held out sterile gowns for each of the doctors then offered up one glove each with the cuff stretched open. As a hand slipped into the glove, each nurse gave a firm downward thrust and let the cuff clamp down on the wristlet with a characteristic snap. Meanwhile, the circulating nurse continued tying the backs of their gowns.

"I'm just grateful working with a living legend like you," Murray added.

"Ha! I wish I were that good, Doctor!"

"You are that good, Doctor!"

"The patient is ready," said the OR nurse.

Dr. Cannon began prepping the patient's skin for surgery, and rearranging sterile drapes to confine the incision site.

AS DOCTORS worked feverishly over their patients at Valley Forge General on that day, November 8, 1944, Chaplain Edward T. Connors sat on the edge of his bunk somewhere in Germany composing the following letter to Bishop Thomas M. O'Leary: *In this region, we find a multitude of devout Catholics. In a number of towns, I was the first American priest that the parish priests had ever met. They have all been very kind and their people faithful. Evidently Hitler failed in rooting out the Catholic faith.*

The commander of the 9th Division is a devout Catholic, General Louis A. Craig. He always receives Holy Communion and is a fine example to the men.

I recently began to instruct one of the battery commanders. He is from the south and had met few Catholics before the Army. He has been impressed by the strong hold the religion has on the men and has asked to be received into the church. "Veritas impellit."

This is something of a consolation to a priest. We do get weary after two years, mostly in combat. A bit weary but never lonesome or discouraged – and I would never think of leaving.

The description he gave of the battery commander closely fit that of a man who became and would remain a life-long friend after the war, General William C. Westmoreland.

THAT EVENING Dr. Murray entered Room 330 and picked up a new patient's chart, everything about the young doctor exuded confidence; straight, tall, lean and handsome, with diamond-edged eyes that missed nothing. You could usually find him in plastic surgery wards jammed with patients, devoting long hours to circulate amongst them, observing then checking closely the results of the most imaginative reconstructive surgical procedures.

Night and day, the doctor wandered through the wards talking to patients, helping with dressings and picking up valuable bits of information. Born with an extreme curiosity, he questioned everything, then questioned the answers, and when he received new answers, questioned them further. Absolutely nothing would he take for granted.

Dr. Murray glanced at the chart of their newest patient then his eyes shot to the bandaged head, as he exclaimed, "Good Lord! Are you the Larry Heron who played for St. Mary's in Milford?" He silently prayed the answer would come back, "No."

"That's me," Heron replied. "Hey, are you...Joe Murray?"

"How could you possibly know?"

"Some people never forget a face. I never forget a voice. You played first base for Milford High, quite the competitor."

"So were you. I will never forget our last game against St. Mary's."

"Back in 1936," Heron said quickly. "You were team captain."

"You have a good memory."

"I will never forget that day. The papers called it a slugfest."

FOUR THOUSAND fans overflowed bleachers and jammed the sidelines watching Murray drive in three runs in the first inning with a single followed by a stolen base. Heron answered with a bases-loaded homer in the second. Murray hit one out of the park in the third to tie the game then followed with a triple in the fifth to put Milford ahead. He would have scored from third base on a hot grounder, except for a perfect throw to home plate by St. Mary's shortstop, Dave Tredeau. The catcher put the tag on Murray's toe as he slid wide, thus ending the inning.

In the seventh, Heron hit a home run with one man on base and two outs to put St. Mary's ahead six to five going into the final inning. That's when Murray came to bat with two out and no one on base.

A hush fell over the stadium as the pitcher wound up. Then a single fan yelled, "Come on Murray. Hit one out of the park." St. Mary's catcher signaled low and inside. The pitch came in slightly outside. He stepped into the ball with a solid crack heard downtown. Murray's homer tied it, six apiece.

Heron's turn came in the bottom of the ninth with two outs. Azelia bit her lip and watched his fingers dance on his cap in a familiar pattern – so quick and subtle that few would recognize the sign of the cross.

"Steer-ike one," the Umpire yelled. Heron swung hard at the next pitch for strike two. Tense moments slipped by as the next three balls in succession passed just outside the strike zone. With a full count, the pitcher sent one down the middle and Heron hit the ball solidly, breaking the bat. What should easily have yielded a homerun fell short because of the broken bat and landed instead between right and centerfield before rolling toward the fence with both fielders in pursuit.

Heron crossed second base as the right fielder reached the ball then rounded third as the ball sailed in a low trajectory toward home plate. Tearing up the turf in long strides as the ball reached home, he slid wide of the plate, barely tagging it with a toe as the catcher swiped at him with his glove but missed cleanly.

As Heron raised his arms triumphantly toward the admiring St.

Mary's crowd, the Milford High catcher walked forlornly behind him and swiped the folds of his shirt in frustration, but Heron never felt it. As far as he knew, the game ended with a win for St. Mary's.

Then why hadn't the umpire signaled him safe?

Heron felt compelled to do something he would later regret. He ambled back and tagged home plate a second time. Why? Because with all the dust and dirt blowing around, he thought perhaps the umpire hadn't seen him tag it the first time.

"Yerrr out!" the umpire shouted, jerking his arm and pointing his thumb to punctuate the call.

St. Mary's fans showered the umpire with boos and blasphemies. Milford High fans, stunned into silence to that moment, started a belatedly low cheer that grew louder as the realization of the reversed outcome of the game slowly spread. The umpire headed for the sidelines after declaring Milford High the winner.

DR. MURRAY'S tone became somber, as he asked, "What did you do to yourself?"

Heron gave a summary of his misadventures, how a shell exploded in his hands and how he went for days unattended and without treatment, finally losing what remained of his eyes in England.

Murray choked back tears, a tragedy magnified by the doctor's personal knowledge of what might have been. "It's time to change your dressings, so let's get started."

"I'm going to give you something to make you comfortable." He swabbed Heron's arm. Taking the IV needle from the nurse, he administered a general anesthetic to prepare Heron for the painful process of removing the gauze. He gently peeled away the greased up gauze soaked with Heron's body fluids, combating shock at the sight of Heron's face, which appeared marked with tell-tale craters and the missing eye sockets covered over and sealed by skin grafts.

The badly scarred face and feral display of teeth formed an incongruous smirk, like some alien creature raised from the lowest depths of the ocean. At least the teeth remained unharmed. The surgeon in England did a fair job of covering the empty sockets with split-thickness grafts, but the ears looked eaten away. Besides the teeth, a healthy crop of brown hair remained his only untouched feature.

Murray read his chart:

Diagnosis:
Burn (white phosphorous) face, forehead, eyes and lids, neck,
and patchy areas of shoulders and hands, third degree, with loss of large
amounts of tissue.
Laceration Wound, severe, multiple, involving lower lip, right
cheek, right anterior shoulder, palms of hands, left knee.
Nasal bones, maxilla, right. A.I. when two phosphorous bombs
exploded in his face as he was loading them.

Operations:
3 July 1944 - Debridement of burns. No anesthesia.
15 July 1944 - Dermatome grafts of face, forehead, jaws,
neck, ears, and enucleation of eyes bilateral. Anes. Endotracheal 10 hours
duration.
3 Aug 1944 - Repair of lip defect. Skin grafts to orbits, nose
& chin. Anesthesia, local procaine.

The doctor could not stem the avalanche of compassion he felt for
this virile athlete, a legend from the same hometown in America where
their high school teams competed. With the loss of both eyes, this young
man's sports days abruptly ended, a flame had died, a bright light extin-
guished.

"I can't believe you're my doctor," said Heron.

"I'm just one of many, and rather new at that. You have the world's
best plastic surgeon in Dr. Bradford Cannon."

"Is he that good?"

"None better."

"Can you tell me a bit more about the man who plans to recon-
struct my face?"

"Sure. I've finished my rounds for the day." Murray pulled up a
chair and began briefing Heron on Cannon, hoping to put his mind at rest.
"I'll start by mentioning his father was Dr. Walter Cannon, the man who
discovered how the esophagus works, and introduced us to the fight or
flight theory."

"The stuff they taught us in school about adrenaline and how athletes get their second wind. Is he the doctor who came up with that?"

"The same. After Harvard Medical School, Bradford did his residency at Barnes in St. Louis, a major plastic surgery center where he first

met Vilray Blair and James Barrett Brown."

"They're the doctors who run that hospital. He told me about them."

"Seven years later at Mass General's plastic surgery clinic, he became a protégé of the world-famous Dr. Varaztad Kazanjian, the man who made plastic surgery a specialty."

"Now I know why you think so highly of Dr. Cannon."

"Up until the early forties, burns were treated with a harsh and painful scrubbing to remove destroyed tissue then sprayed with tannic acid to form a protective membrane. Dr. Cannon and Dr. Oliver Cope found the membrane forestalled healing and led to infections, so they began covering wounds with fine mesh gauze impregnated with petroleum jelly, which proved less invasive and not so painful, but that it prevents infections and speeds up healing. Fortunately, their findings occurred just prior to the Coconut Grove fire."

"Were you practicing then?"

"I was in grad school, but it's as though it happened yesterday. November 28, 1942. It was the night of championship football games and to celebrate Holy Cross upsetting Boston College, partiers crowded into the Coconut Grove nightclub. A domed ceiling of decorative fabric ignited when a busboy lit a match to see better as he changed a Christmas bulb."

"Dr. Cannon left his turkey dinner to return to Mass General to treat a hundred and fourteen of the most severely injured; all but thirty-nine died within hours. Thirty-six of the hundred thirty-two taken to Boston City Hospital died as well."

"Bet that's when their new treatment withstood a first real test."

"Exactly, but enough about us, let's hear about you. I don't know how anyone could have survived what you experienced."

"I had to get back to Azelia. Love her too much to just give it up."

"I know what you mean. I feel the same way about my fiancée."

"She must be someone very special. How'd you meet her?"

"My last year in med school, I attended a Boston Symphony Orchestra concert with some classmates and spotted an incredibly lovely young lady I thought far too nice for the fellow she was with."

Heron laughed.

"'Accidently' finding her in a corridor alone, I learned her name was Bobby Link, a music student majoring in voice and piano. By intermission's end, I knew we would marry."

"So when's the big day?"

"This June."

"Congratulations. Sounds like we were both lucky in love."

Heron with fellow patients at VFGH in 1944.
Courtesy of Larry Heron, Jr.

AZELIA FEELS at peace, content in the knowledge their love kept him alive in France, and contributed to his ability to endure pain during the scrapping of those terrible wounds, operations one after another, and long hours of lying unmoving in one position, patiently waiting for cells to form, die and rebuild. And she feels thankful too that God brought Doctors Murray and Cannon into his world. He timed lives and interwove futures so all paths crossed at exactly the right time, not only to provide proper skills needed to repair her husband but to bring comfort and understanding to his bedside.

First Father Connors ran into Heron on the battlefield then Doctor Murray came across him in the hospital. Coincidence? Doubtful. God felt the need to make up for an awful mistake by bringing them together as He did. Azelia does not believe in coincidences but she does believe strongly in Almighty God.

So relaxed does she become while reflecting on the strength of their love, that she lets her race with the moon slip from her mind and nearly dozes off, but only for a brief moment, soon to get back on track.

MURRAY STOPPED by to visit with his old friend as often as time allowed. The doctor took a personal interest in his progress, elevating Heron's spirits whenever he found time to drop by and reflect on better times.

Dr. Cannon also stopped to check on his new patient often. On one such visit, he said, "I hear you and Dr. Murray were baseball rivals."

"I once read he could have played professionally, but it seems his first priority was to become a top surgeon."

"Quite true. Joe was one of only two from his class at Holy Cross that made it into Harvard Med School."

The two doctors obviously shared great respect and admiration for one another, and without a doubt, both tops in their respective fields. Heron felt grateful he was sent to Valley Forge.

On Monday, November 13, 1944, the temperature outside Valley Forge Hospital in Phoenixville dropped to 40 degrees, a day when Heron underwent his first major surgery on U.S. soil under the capable hands of Dr. Bradford Cannon.

First, the doctor released scars around the corners of his mouth to enlarge the surface area of skin covering his face. He did this to allow the underlying muscles and nerves to function so that Heron could open and close his mouth with relative ease.

This procedure would be repeated in stages during subsequent operations. Since burn scaring contracts the skin in every direction, Cannon incised the scars on both sides of his face from the corners of his mouth upward towards his ears. Heron's scars reacted like elastic bands when cut, with the freed ends retracting to form craters of raw tissue which required skin from Heron's flank and thighs to fill in and smooth over. The last step in the procedure called for skin grafts over the remaining burned areas of Heron's face and lips.

That night he lay swathed in bandages, his world reduced to a small bed in the corner of a tiny room, a world forever masked by an eternal shroud of darkness as black as the center of a Pharaoh's tomb, but in the end, everything went exactly as Dr. Cannon previously described. The burned areas scraped raw in preparation for new skin grafts received matching skin harvested from healthy parts of his body, another painful procedure that kept Heron on the verge of screaming despite the administration of strong sedatives. After the operations, dressings protecting the new skin required replacement on a daily basis until new growth appeared, and the dangers of infection were passed.

On the morning of November 16, an orderly stopped by to read parts of a newspaper to Heron, who preferred listening to the headlines first so he could keep up with the latest news on activity at the front. Usually the sports pages would come next but today Heron stopped the orderly from proceeding after hearing an article regarding heavy fighting in the progress of the war, saying wistfully, "I'll bet the 87th is in the thick of things, right up there in the front lines beside the infantry."

CHAPTER SIXTEEN

A NOT SO MERRY CHRISTMAS

JUST AS Heron imagined, that precise moment found Company A in the vicinity of Birgel, attached to the 331st Infantry, 83rd Division, and Heron's platoon now led by Sgt. Nicola. That morning they began firing high explosive and white phosphorous rounds at enemy positions near Krauthausen, Germany.

During a lull in the action, John Sears pulled up in a jeep and immediately searched out Cookie for some coffee and snacks. Prior to the landing, he'd been transferred to Battalion Headquarters and reassigned as Brigadier General Barber's driver. The jeep he drove up in bore a plate on the front that held one star. Upon taking on his new assignment, he chose as his first order of business to remove the jeep's canvas top to improve peripheral visibility then take out the windshield so as not to reflect the sun and lastly, wrapping the front plate with olive-drab material to hide the star from enemy view.

"So what's new," Cookie inquired. Somehow, Army cooks always manage to insert themselves as intricate cogs in the rumor mill.

"All hell broke loose in the Ardennes," Sears replied. "The bloody Krauts are pouring through by the tens of thousands in a straight line for Paris."

"Ar-dens?"

"It's a huge forest that runs through eastern Belgium and northern Luxembourg."

"Geez, the man's a geography whiz as well as a historian," one of the chefs piped up.

"When they reach the Siegfried Line, they're going to get their butts kicked," Cookie admonished. "The news coming in with reserve troops says the war in Europe is as good as over."

"Tell that to the Krauts," Sears stated flatly. "Not only are they on the offensive, I hear they plan to drop paratroopers behind our lines. Hitler's hell-bent on retaking Aachen."

"You're Barber's driver aren't you? You must be in the know."

"I didn't get that from Barber. They could be headed this way is all I'm saying. Just keep an eye open for them

"They've got to go through us to get to Aachen," one of the helpers speculated.

"You got that right," Sears shot back. "Right through us, and we aren't about to let that happen."

"Hope you're right."

SEARS' REMARKS proved accurate. On December 16, Hitler's Belgium troops crossed the Meuse River to unleash his Fifteenth Army on Aachen then followed up with a massive offensive against a weakly defended American line in the Ardennes. Caught by surprise, the line collapsed and the Germans smashed a forty-five mile hole through which German armored columns swept to attack supply dumps located along the Meuse River.

On December 17, in the vicinity of Kufferath, Germany, with the temperature falling below freezing and an icy wind gusting to over seventy-five miles an hour, Pvt. Angelo Bastoni and First Lieutenant Marvin J. Dewitt acting as forward observers, began directing fire on enemy positions from a house on the high ground in Burzbuin, about 2000 meters northeast of Gey.

Bastoni held his nose to ward off the acrid smell of rotting flesh wafting from piles of dead horses the German army used as a principal means of transportation due to lack of fuel. The stench became as pungent as acid and thick as smoke, and he could not block it out with a sleeve or even a loose rag clamped over his nose.

Suddenly, he heard the dreaded grinding of tractor treads. Worse still, the noise abruptly ended. Did the Germans spot the Americans? Was the tank about to fire at them? Both answers came a second later with a

"Whump!" that separated Bastoni from the ground and rattled his brain. *Whump! Whump! Whump!* Three more explosions followed, one that threw Marvin Dewitt twenty feet to lay on his back with blood pouring from his nose, ears, and mouth. Bastoni knew instantly the lieutenant was dead but checked for a pulse nonetheless. Seeing no external wounds, he supposed the concussion killed him. Time to get the hell out! *Whump! Whump! Whump! Whump!*

THE BUILDING stood in a small glen, hidden by a thicket about a mile from where he left the lieutenant's body; he had made mental note of his way back to retrieve it later. To his surprise, the structure remained intact despite the loss of doors and windows from the blasts by now moving further and further away. He strained to listen to ascertain no enemy troops waited inside then cautiously entered through a blown-out side door. He found it slightly warmer inside, the brick walls blocking out the wind but not the cold.

His body reminded him suddenly that he had not slept or eaten a thing in over seventy-two hours. Darkness fell quickly and after listening to the extreme quiet for five straight minutes, located a corner behind an old wardrobe closet in a corner of a bedroom to curl up in a fetal position and catch a little sleep. The next thing he recalled was waking from a dream which found him back home on Christmas Day with his mother standing beside the Christmas tree looking at him with tears brimming her eyes.

As he came fully awake, he blinked involuntarily to dispel a gritty substance from his lips, and experienced the feeling that an elephant squatted on his chest during the night. He tried to move but the elephant held him in place, a heavy weight pressing down on him. Powerful sunlight poured in, around, and over him.

Morning already.

Bastoni heard them speaking German before spotting them framed in an opening where the front entrance stood the night before, three armed enemy soldiers just twenty feet away. Two looked much younger than the third and lacked whiskers between them to make one penciled mustache. Their leader moved like a man of forty as he looked over at Bastoni with the tired eyes of a man twice that age sporting iron-gray hair from under his helmet and a splotchy gray beard.

Sometime during the night the enemy must have brought up tanks.

The building obviously took several direct hits and lay crumbled around him. Half the roof came down through the second floor, landing on his right, which explained the bright sunshine. Good Lord, he'd slept right through a barrage and tons of debris and bricks that fell around and on top of him! With the onset of daylight, they came to investigate. Still groggy, he had yet to move a muscle. Thank God for that. His head remained tilted back at an angle that allowed him to see through his lashes without showing signs of life.

The tip of his rifle poked out from under the ruble and beneath it his hand came in contact with the stock. In the time it would take to free it, they could unload their weapons at him. Fine dirt particles lay like snowflakes over the surface of the jagged detritus, some of it tickling his nose so that he desperately needed to sneeze, but to do so meant death. Each of the Germans gave him a cursory glance but seemed unconcerned. Who would have lain still through the night with dirt and bricks piled up on him if still alive?

Then he heard a noise from somewhere behind him. They heard it too. Six eyes focused on him like gun barrels and rifles aimed as he prepared to die. Suddenly, the Germans broke into nervous laughter as one of the younger men picked his way toward Bastoni. *Are they laughing at me?* Just when he thought the jig was up, the young German vanished from peripheral view, while sweat made a path down Bastoni's neck. Seconds passed then the soldier came back into view holding a scrawny kitten, scratching it behind the ears. Bastoni could hear it purring like a motor, which seemed to put the Germans at ease.

After the enemy soldiers cleared out taking the kitten with them, he waited ten minutes then freed himself and his weapon from the pile and proceeded to shake off the dust. An hour later he made it back across friendly lines, and a half-hour later lead a search party to retrieve Dewitt's body.

THUS CAME a turning point in the war. General Von Rundstedt's forces launched a counter-offensive that broke through a weak spot in the American lines to start the campaign noted by history books as the *Battle of the Bulge.* Deep snow, severe cold, and intermittent rain would render this one of the toughest campaigns of the war, with Company A at its forefront.

The battle began on December 16, 1944, ending 31 days later on January 31, 1945 when Allied forces united to close the Ardenes. The

Germans then launched massive artillery and three armies at the Americans. Due to dense forests, no fighter planes took part but 700 American tanks went up against 500 German Panzers. When it ended, the entire 87th Battalion pulled out of Germany and into Belgium to take part in a stand that would turn the tide of the war. One small section of Company A moved from Schleckheim at 0900 hours, arriving in Havelange at 1400 hours, while the main body jumped off from Ober Forstbach to bivouac in a small town near Marche.

Battle of the Bulge, December 16, 1944 - January 15, 1945.
National Archives photo, courtesy of The World War II Database

During the height of the coming battle, Christmas Eve would fall quiet on December 25 affording the men the opportunity to sit down to their first Christmas dinner overseas. A splendid dinner prepared by hardworking kitchen staffs helped offset the depression and fatigue that had been building throughout the terrible fighting preceding it. The menu that day featured turkey, potatoes, vegetables, cranberry sauce, pies, candy, fruit, and a few rounds of confiscated liquor.

CHAPTER SEVENTEEN

RECONSTRUCTION
AND
REHABILITATION

THAT CHRISTMAS, Azelia took a plane to Philadelphia and a room at a nearby hotel to spend the holidays with her husband. They enjoyed sleigh rides and dances, and joined in singing Christmas carols. At a concert held on Sunday, December 17, Heron sang *White Christmas, Silent Night*, and *What Child is This* to a packed house.

His voice sounded soft yet riveting, powerful yet rich, but not as shrill as an Irish tenor's. The audience stood captivated by more than just his voice. Standing ramrod straight with his cane characteristically held at a slight angle from the ground, he appeared much taller than his six-foot one. He finished to a standing ovation, with the crowd expressing admiration, not pity, though enough tears fell that day to fill Lake Erie.

Later, a fellow patient asked, "How can you be so elated? Every Sunday when you show up at Mass and sing in the choir, you lift my spirits and make me ashamed for feeling sorry for myself. Don't you feel any bitterness or anger towards God for what you've endured?"

To that he replied, "Lots of good people died in this war. I'm one of the lucky ones. The only thing I regret is not being able to fight alongside the 87th at this very second."

On the 25th, the Herons sat down to Christmas dinner in the Valley Forge Hospital mess hall. Surprisingly, the food tasted great, including a rich mélange of turkey, mashed potatoes, gravy, butternut squash, corn, and cranberry sauce. After he finished everything on his plate, Heron patted his swollen stomach and felt more relaxed than at any time since receipt of his injuries. Azelia's presence made a vast difference, elevating him to a good frame of mind.

"Like another helping?" she asked.

"Soon as I burp."

"Larry!" she scolded.

"Well, I've got to make room because I definitely want seconds."

After dinner, they sat drinking hot-mulled cider and Azelia filled Heron in on the news back home. She read the names of locals listed as killed, wounded, or missing in action from back issues of the Milford Daily News and the Worcester Gazette. An end to the war could not come soon enough.

ON MONDAY, January 22, 1945, Dr. Cannon grafted more skin to Heron's face and chin. On March 6, he added grafts to Heron's cheeks and neck with skin tissue taken from his back, buttocks, thighs and abdomen, once again making it impossible for him to find a comfortable position to lie in.

On April 16, the doctor explained what to expect next. "We'll smooth out your face a bit and rebuild your left ear with cartilage from your rib cage. At the same time, we'll prepare a tissue flap roughly four by seven inches long from your upper arm, called a pedicle graft."

How interesting. The bile rose in his stomach at the thought of yet another portion of his body mutilated.

"Why the arm?"

The doctor explained how one end of the flap must remain connected to his arm until a nose hookup becomes established. Once it builds its own independent vascular system, the flap could then be permanently disconnected from the arm.

"Is this a new technique?"

"Italian rhinoplasty has been around since 1597."

"A nose from arm muscle."

Suddenly, the door swung open and a nurse entered carrying a tray.

"Oh. Sorry doctor," she exclaimed. "I didn't mean to interrupt."

"No problem," Dr. Cannon said. "Take your time and do whatever is needed."

"I'll be out of here in a second." Her shoes sounded *clop, clop, clop,* as she approached the opposite side of Heron's bed from the doctor.

"Hello, Jane," Heron said.

"Good morning." That Heron could identify her walk came as no surprise, nor the sound of her voice.

Heron heard pills falling into a small container, the container placed on the tray, water spilling into a glass, and the occasional clunk of ice. "You need to take these pills now," Jane Barsanti said.

After she left, Dr. Cannon asked, "To prepare the nose flap, we'll have to set a proper thickness, de-bulk where necessary and plan ahead for contraction. To that end I will cut the flap larger than the area it will cover to compensate for shrinkage during healing, and for the later effects of atrophy and gravity."

"How long before you detach the flap from my arm?"

The doctor felt it best to set expectations to remove any uncertainty and apprehension. "It usually takes about three weeks."

"Three weeks?" Heron gasped. "That's a long time."

We'll strap your arm to your head and brace it against the head-board to prevent the flap from premature detachment. During this period, we will continue to irrigate the wound-bed to keep it from drying out and applying suction to evacuate fluid accumulation. At all times, we must insure the flap remains firmly pressed against the wound-bed."

The prospect of his arm locked in one position for such a lengthy time would force him to rely on nurses for everything: bedpans, baths, clothing changes, even scratching. "How will you know when the flap develops its own blood supply?"

"We'll test for it with a flourescein dye introduced into the blood-stream intravenously."

"Okay doc." Heron heard enough. From the day of the injury, he knew nothing but pain, circles of pain, ripples of pain, slamming unre-lenting pain. Every night he drifted to sleep with it, dreamed of it then awakened each morning with it still present. Now he must look forward to another extended period of prolonged pain and discomfort. But he must not, will not give up. He gave her his promise to see it through. Azelia had his word. Perhaps after another year or more of operations he will wake up one day with a real face, one that she could stand to look at. Azelia de-served that much.

ON MAY 8, 1945, Azelia sat in the waiting room with newspapers piled beside her. Bold headlines declared Victory in Europe; history would later record it as VE Day. *Thank God.* But Larry's war continued, in fact, the battle to get his life back was just getting underway. He'd been under the knife since early morning with no end in sight.

Heron following extensive plastic surgery.
Courtesy of Larry Heron, Jr.

Dr. Cannon came out to see her at noon looking like a man in need of a good night's rest, still dressed in scrubs with spots of blood near his lapel. "The operation went well," he said with a smile. "The skin grafts and cartilage transfer to the ear went exactly as planned. Also, the arm flap has been prepared, which will serve to improve his nose."

"Flap?"

Just as with Heron, he gave a quick overview on the subject of rhinoplasty. When finished, she pictured the arm muscle carved like a turkey breast, one end of the slice left attached to the arm to keep it 'alive.' Her knees grew weak and her stomach queasy, and she wondered if she could ever look at turkey breast as she once did. But it elated her to see Heron up and moving about later that same day, although it bothered her to see his face completely bandaged, as well as his left arm. And his hip and thigh bore more bandages to cover places where the doctor harvested tissue. In addition, heavy lengths of cloth bandages crisscrossed his chest and back like bandoliers and his bare arms showed scars from shrapnel wounds, and it especially pained her to look at the bandage protecting his arm flap.

Heron asked her to slip his sunglasses on over bandages. When she complied he asked, "How do I look to you now?"

"You don't really want to know," she laughed.

"Funny? Do I look funny?"

"The opposite. Kind of scary. Do you remember the H.G. Wells movie we saw, *The Invisible Man?* "

Heron tried to smile. "I look like Claude Rains? Everybody ran from him and obeyed his every command. Are you afraid of me?"

"That'll be the day," she laughed, sardonically then asked, "Do you know what they plan to do next?"

"Yeah," he said blithely. "Build me a new nose."

B Y J U N E 5 , the grafts healed so that Dr. Cannon could add new grafts to both ears. At the same time he reshaped the arm flap for later nose reconstruction.

From the beginning Heron fought becoming an addict by insisting on no more than four morphine shots a day, two less than his caregivers wanted to administer, preferring to deal with pain rather than addiction.

By June 15, he felt healthy enough to proceed with skills training in anticipation of returning home for a short spell, during which time the latest grafts could heal completely. Though the wounds to his face no longer required bandages, reconstruction would continue for years, his face would remain a patchwork and his ears still required more repair. The arm flap showed signs of healing but would remain bandaged to protect it from infection.

It concerned him that it would take so long to repair his nose and that his left cheek felt gouged from shrapnel wounds, and still had craters. When he ran his fingertips over his face, it felt like a living portrait of Dorian Gray. Others he felt, must look upon his face as made of wax and left too long under a hot sun.

"You'll be going home next week," Nurse Barsanti told him one morning.

By this time Heron had changed his mind about returning home so soon. "I don't want to," he stubbornly insisted. "I prefer staying right here until this is over."

"What on earth do you mean? I thought you looked forward to returning home to visit friends in familiar surroundings."

She had experienced it with her patients many times. They would balk at leaving their comfort zones, fearing they might shock friends and relations with their change in appearance. They felt ashamed because they looked so different from the time they left home. Anticipating their gasps and stares struck them terrifying prospects.

As if to confirm her thoughts, he said, "I'm too ugly and scary right now. I'd rather wait until I look more normal."

"Nonsense," she said, knowing he would like to put it off indefinitely.

"I am ugly."

"Okay. Let's say you have yet to reach the handsome prince stage. I know just the thing that'll solve your problem."

"What?"

Just then an alarm sounded.

Nurse Barsanti let out a sigh of relief. "Emergency!" she shouted heading for the door. "Gotta go. I'll have it for you on Monday."

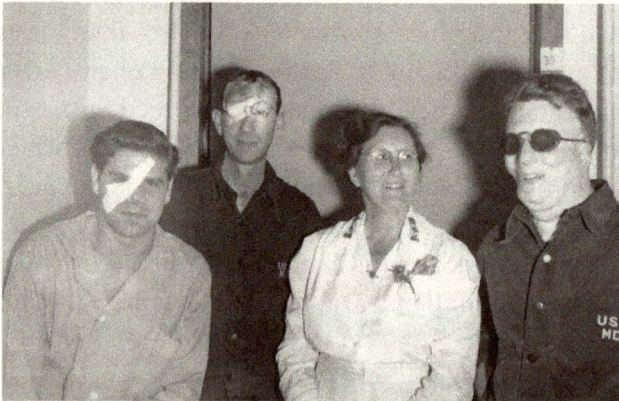

Larry Heron with VFGH Army nurse and fellow patients.
Courtesy of Larry Heron, Jr.

TWO TEENS tossing a baseball back and forth on the Community House lawn stopped when they saw him approach. "Yikes! Look at that!" one boy said, slowly backing away as the spooky guy approached.

Invisible Man and *Frankenstein* movies reruns at the State Theater
fired their young imaginations, and this man wearing sunglasses over a
long black mask looked even scarier than the movie characters.

"I know who that is," his small friend said. "He's that soldier who
threw himself on a grenade to save his buddies."

"Wow! I'd never do anything like that."

"Me neither."

EVERY STREET he walked, every building he passed held a memory
and every person he chanced upon had a story to recall. As he painstakingly mapped streets and key locations in his mind, he unlocked a trea-
sure-trove of memories. The years rolled back as he made his way through
town and his mind raced back to a kinder, gentler time. The most pleasant
memories involved his courtship of Azelia, the person to whom he owed
his life. She gave him a reason to live and now that they were reunited
made life worth living, or at least the most important aspects of it.

The town remained much as when he left it, homes owned by
the same people who now greeted him warmly. Though his closest male
friends remained at war, nothing fell to neglect in their absence. Despite
the paucity of manpower, nothing in town went unattended. Women and
children of all ages stepped up to take responsibility for the important
chores and tasks previously handled by men. The sidewalks stood clear
of kid's toys, trash, or obstacles that might impair his progress, while the
smell of fresh-cut lawns and carefully tended flowers gave proof of proper
care.

On Sundays, people greeted him at church and those out for a stroll
stopped to chat while still others called out greetings from porch swings
or front lawns. Heron astounded people by recognizing their voices and
wherever he walked tales of his heroics on the battlefield preceded him.
"There goes the man who lost his sight and gave up his future to save so
many lives."

Not many in Hopedale and surrounding towns remained unaware
of his heroics and the sacrifices he made, or of the beautiful young wife
totally devoted to him. Plenty of men still sought to test her resolve in
subtle and not so subtle ways. Why would she want to spend her life tied
to a blind man who could not adequately provide for her, especially since
she could find a far easier life with most anyone of her choosing, a man

in good health with a sound future? Azelia made clear the answer in the way she dressed and carried herself. Make no mistake, she would forever remain "a one-man woman."

Those who knew their story took the couple into their hearts and admired Azelia as much as Heron for her outstanding courage and devotion. Her steadfast loyalty and acceptance set an example for others, and tore down fences those seeing Heron for the first time might have erected. The battlefield served as a sort of proving ground for those who fought and returned, sparing none of the returnees, all seemed different in some respect from when they left. If those who made it home alive showed no obvious physical wounds, they most likely suffered some degree of psychological damage.

Heron left home a sports figure with a brilliant future and returned to his hometown a blind man who struggled now simply to make it safely across a street. But for all he suffered, he never lost his sense of worth, and certainly maintained a strong sense of humor. Instead of becoming reclusive, as some imagined, Heron remained as outgoing as the day he left, shunning pity or sympathy. Always resourceful, he insisted on handling everything for himself, asking help from no one, and because of his fierce independence, commanded the same respect and admiration as the Heron of old, the sighted athlete. Friends and neighbors stopped regarding him as a blind man and instead saw him as a man who could match wits, who walked with a bearing and commanded total respect. Everyone wanted to be counted as his friend, and most wanted to know what he hid behind the black mask, but did not dare ask.

Heron enjoyed his walks about town, a solitary figure stopping to take in his immediate surroundings by sound and smell, one sip at a time. Deprived of the pleasures of looking with a pair of eyes upon the exquisite beauty of the town, his mind would seek to recapture and play back old images. On his first walk, he stopped in front of the centrally located Bancroft Memorial Library, erected at the corner of Freedom and Hopedale by Joseph B. Bancroft, who in 1898 donated it to the town in memory of his wife, Silvia.

Copied after the Merton College Chapel at Oxford University, in its day the structure was voted the state's most beautiful library. Inside, the familiar smell of books and the quiet that descended upon the solidly built structure remained exactly as he remembered. Six cathedral-styled arches spanned the width of the main reading room above walls adorned with rich oak-paneled wainscoting and carved moldings. In answer to his query, the librarian told him the walls above the green-glazed tile fireplace and mantel of oak still displayed the portraits of Joseph and Silvia Bancroft.

He walked down a flight of stairs and out through the door in the children's section, and as he moved around the building, passed his fingers over a dolphin, a pair of eagles, Medusa and parts of the fountain beside the Roman goddess named Hope. All had been carved from Carrara marble and placed outside the building by Bancroft as a symbol of eternal goodness.

On subsequent walks along Hope Street, he would turn right onto Hopedale Street and proceed the quarter-mile past the all-brick Draper factory, a building he remembered with ribbons of mullioned windows running its entire circumference. Where Hopedale crossed Freedom Street, he could turn left to cross the bridge spanning the waterfall used to generate hydroelectric power for the acres of machinery within, then on the other side turn right to trace the worn path along the opposite side of the pond where he often went to fish.

Instead of crossing the bridge over the waterfall, he could continue on Hopedale Street, following the road as it curved past the pond to where he would cross Dutcher Street and pass through the entrance one of the best equipped parks in the country. In the park he would hear children playing and adults whacking tennis balls on one or more of its three clay courts.

Yet another route took him through the tiny business district, and another led past the cemetery, or might take him along Adin Street, with its high-walled fortress-like surroundings that hid the Draper mansions from the view of curious passersby. And when he tired of Hopedale, he would make his way around the neighboring town of Milford.

"I ran into Larry Heron at the Community House this morning," Bill Larson told his father one day. "As I opened the door on my way out of the back exit and started down the stairs, I spotted him coming up the walk towards me. He froze when he heard the door open, and I froze, too. He probably thought some idiot stood staring down at him instead of holding open the door."

"Did one?"

"Go ahead," I told him. "He wouldn't move. I knew then that he didn't want my help, just a clear path, so I let the door close and continued down the stairs. When I looked back, I saw him enter on his own."

Once Heron made it clear he did not view himself as handicapped, the townspeople quickly learned to treat him accordingly.

"There's plenty to watch out for," he told Azelia. "Vehicles moving at incredible speeds, kids whizzing by on bikes, cars not always parked where they should. I feel vulnerable at times but I'll tell you, I am not afraid out there. Fear has no place in my world." She resisted responding with, *You were never afraid of anything. Perhaps if you had, you'd be fine today.*

He hadn't been away so long that he couldn't visualize the grandstand in the park where band concerts took place on Wednesday nights, the bench by the pond where he made up with Azelia after their one major spat. He could still hear the sound of water lapping the shore by the pond on that gloomy sunshiny day. The memories remained, yet it seemed strange to pass familiar places and depend on his imagination to "view" them.

Whenever he left the house, he would wear the black mask that Nurse Barsanti presented him with before he left the hospital. Heron put it on and refused to take it off even in Azelia's presence. She detested the phantom aspects of the mask and wished she could steal it from him and dispose of it but instead recommended he stop wearing it. "People's imaginations substitute images more dreadful than the face you desperately hope to conceal."

"You have no clue as to what's under this mask," he told her. "If I remove it, you'll get a jolt of reality that will shock you."

"I've already seen you without it in the hospital."

"Okay. But seeing it once or twice was enough. You don't need a reminder."

"Do you really think your scars matter to me?" she asked.

"They will, when you see them again."

"Then you don't know the woman you married."

No matter how bad his face might look to her now, she knew she would eventually become inured. It bothered her more that he returned with any injuries at all, and she dreaded the painful operations that lay ahead for him. She would lie awake for hours at night thinking how much the explosions hurt and terrified him, imagining him writhing on the ground, screaming in pain, blinded and so terribly far away from her and so horribly alone.

To help speed Heron through the healing process she must first convince him that looks made no difference and that he still remained her perfect man. She could think of only one sure way to accomplish that goal. The answer came as she lay soaking in the bath tub one day, as did many good ideas and solutions to problems. The next day, she finished her bath then refilled the bathtub and led him to it. Lifting the mask high enough to slip him a warm kiss that elicited a gasp. She then told him, "Meet me in the bedroom when you're finished, darling."

When he finished his bath, she guided him to the bed and removed the towel wrapped around his waist. Then she propped him against a pillow and climbed in beside him.

"Now take off the mask," she commanded.

"You'll want me to put it back the minute you see my face," he said self-deprecatingly.

Azelia remained silent.

Heron cleared his throat. "There's something I've been meaning to ask."

She closed the distance between them so that they touched the length of their bodies and the sweet scent of her caused him to tremble. Then, lifting his hand in both of hers, she raised it to her lips to kiss first the back and then the palm.

It seemed the clock had been turned back and they were on their honeymoon. He pretended that instead of being blind, he merely had closed his eyes. She caressed him with fingertips warm, soft, and inviting. He touched her and the familiar stirrings began. It had been so long since they made love that his knees felt weak, yet a lingering uneasiness remained that he couldn't explain.

He swallowed hard. "Did you know that 1945 is going down as one of the best years for Bordeaux?" he asked.

"Is that what you wanted to ask me?" she laughed.

"Not exactly. I want to know how you feel about having a child. I mean, this might also be a good year to start a family."

Her long-term plans included children but on this day could not be further from mind. Babies require constant attention. It would mean dividing her time between raising a child and caring for him. She might be forced to quit her job right when the need for extra cash became paramount. Of course they would still have Emma living with them. With her help Azelia could be back at her job within a few months.

And in his present state of mind, the slightest hint of rejection could defeat her purpose, so she answered, "Great minds think alike." A moment passed before she said, "Perhaps one child."

"You shouldn't look at me. Might turn you off..."

"Just be quiet," she ordered. "Now off with that damned mask!"

Gently, lovingly, she lifted off his sunglasses and placed them on the night table. He held her hands for a moment then passively removed them as she slowly slipped off his mask. At the sight of his face she bit down on her lip, stifling a gasp. The monstrous patchwork without eyes and a twisted smile shocked her, just as he had warned Closing her eyes, she tried to picture him in her mind before the injury, without the scars, without any missing pieces and multiple skin grafts. She visualized his blue eyes first then his jaunty chin, his sensuous lips and cocky grin, the first and last man she ever loved. He could still jump-start her heart with his touch. Thank God she had him back in her arms once more.

The whisper of an empty robe sliding off the bed and hitting the floor sent his pulse soaring.

"My God she said, touching him tenderly, you're more beautiful than I remembered." She bent forward to kiss him gently on the lips, forehead, and cheeks; touching each scar on his neck, shoulders, forearms, hands, thighs, legs, and feet. She touched his chest with her lips and he shuddered involuntarily.

Rivers of emotion swelled inside his body spilling their banks, spreading, building, gathering momentum. Each kiss was delivered lovingly and with the utmost tenderness. He divined such bliss that tears, finding no outlet on his face, made a path down his nostrils and into his throat, and then he tasted her tears.

She kissed his lips lightly at first then longer, deeper. When they finally made love, it was like a dream in which they melded one body, one soul, one kernel of life suspended in time, neither uttering a sound. Just one word would have upset the balance, destroyed the mood - changed everything.

She wished her love could heal his wounds and restore his sight, that time could be reversed and make him whole again. But she would settle for letting him know once and for all that he was loved by her, and to that end, she would devote each day of the rest of her life.

CHAPTER EIGHTEEN

THE LAWRENCE J. HERON POST

ON MONDAY, July 9, Azelia drove him to the Providence train station after a heated discussion. When she finished her lengthy discourse on the one hundred-and-one ways something bad could happen to him, he took her in his arms. "Look. Sooner or later, I'm going to have to make my own way. You driving me to Philadelphia will only hold me back. I want to learn to travel to and from the hospital on my own."

"But you could get lost."

"Not possible. I get on the train in Providence and off in Philadelphia. Ralph Davis will meet me at the station and take me to the hospital. "

"But you'll have to change trains."

"We talked about this. They pay people to see that I get on the right one. Nothing could be simpler."

She guided him to an empty seat and placed his bag by his feet. "Call when you get there," she said through tears streaming down her face.

"I will."

She gave him a long kiss goodbye then held a handkerchief to her eyes, for the time arrived for her to leave. Three soldiers boarding behind them waited patiently for the aisle to clear then one of the men asked Heron, "Mind if we sit with you?"

"Not at all," he answered.

"Don't worry, ma'am," the soldier said with a friendly drawl, "We'll see that he gets to where he's going."

"You got somebody on the other end?" he asked Heron.

"Yes. An orderly will be waiting to take me to the Valley Forge Hospital."

"Shoot. That's where I'm headed. I work there. What's his name?"

"Ralph Davis."

"I know Ralphie. Ma'am, like I say, you don't have a thing to worry about."

Azelia felt a great sense of relief leaving Heron in the company of three obviously fine young men who promised to look out for and protect him, safer with them than in her car.

On the day Azelia accepted him without the mask, Heron dispensed with it, though he remained aware his face would draw stares. There followed gasps, even cries from frightened children, but he could live with that. These soldiers worked at the hospital where they encountered many wounded men, so his looks did not frighten them. He felt in good hands.

A Z E L I A   W O U L D not see him again until Labor Day weekend when she arrived wearing a bright smile and emitting a glow he could not see. But he did sense happiness in her voice when she suggested, "It's such a gorgeous day. Why don't we go for a walk?"

She located a bench in the shade of a tall oak and once they took seats, gently lifted his hand in both hers, placing them on her stomach. "Feel anything?" she asked.

"A lovely tummy."

"Anything else?"

Heron jumped to his feet. "Are you...are you telling me that I'm, that we're..."

"You are going to become a dad," she blurted out with laughter and joy.

Heron leaned forward to hug her then sat down and snuggled close. "You shouldn't have driven all this way by yourself. You should be home resting." She heard genuine concern in his voice, while his mind's eye formed an image of her beatific face as she cradled their child lovingly in her arms.

"I'm not alone," she said. Your mom and Ethel came with me. They decided to wait in the reception area until I gave you the news."

The news of a child on the way affected his spirits like an elixir from the gods. Suddenly he felt ready to face any amount of pain and suffering promised by multiple surgeries planned over the next two years and longer, and though he knew he must reach deep inside for the fortitude to survive the prolonged ordeal of facial reconstruction, in nine months he could look forward to the delivery of a bundle of joy.

THE MOON has nearly reached the three quarter point in its trajectory across the broad window pane, as she recalls Heron entering a new phase of depravity and unadulterated pain visited upon him. No man should suffer as badly or for so long as he did for most of his life. To have been born with gifts beyond the dreams of most men only to have them snatched from his grasp at the peak of his prime seems a bitter travesty, yet he accepted his fate and rose above it as often as he could, but these next few months would test him as much or more, adding new meaning to the words "pain" and "deprivation."

DURING a lengthy operation on September 6, Dr. Cannon attached the loose end of the flap to Heron's nose, leaving the other end anchored to his raised arm. To maintain the integrity of the connection, nurses lashed his arm to his head and secured his wrist to the headboard of his bed. This was the worst form of torture, like being buried alive. This surely pushed the limits to the amount of pain and discomfort any person could endure. Shots of morphine helped, but nothing short of death could match his current misery, and to make matters worse, the migraines began with no signs of let up.

Nurses and orderlies administered to his every need, not allowing him to move his head or arm independent of one another. Following a week of torture, he asked Dr. Cannon, "How much longer?"

"Just a few more weeks. I wish I could say days, but time is a critical factor. It's taking nicely though, no infection. Other than the discomfort, how's everything else feeling?"

"It's a little tough to eat and breathe with an arm and bandages stuffed in my mouth. Other than that, I'm doing just fine." His voice sounded muffled.

"I'm glad to see you kept your sense of humor. You know, anyone else would complain endlessly."

"Just tell me when it's time for me to move again."

The doctor laughed. "That's not a complaint, is it?"

"I don't think so but if you want to hear a complaint, I am sure I can accommodate you. For example, every part of my body coming in contact with this bed has developed new sores."

"I'm really sorry for the inconvenience. We're doing everything we can to help you through this. I wish we could make time pass faster."

"I know. And I would not have made it this far without the help of the nurses. They've been wonderful. I could not do for someone else, the things they do for me while I am glued to this bed. God bless them. They make it so that I don't even feel embarrassed anymore."

"Let me know if we can do more for you."

"A shot of Jack Daniels would help."

"Sure. Do you take it with or without morphine?"

The next two weeks dragged by like years and finally on September 27 the doctor detached the flap from the arm, leaving Heron with a new nose. When he came out of the anesthesia he found that though his arm had been detached from his face, it wanted to remain in the same position. "Nurse, I can't move my arm."

Nurse Barsanti said, "It's been immobilized for too long. Let me help you loosen it a bit." She massaged his arm very gently at first, moving it a few degrees at a time. It took several days of massage and therapy before movement came anywhere close to normal.

But respite from pain and torture must wait.

Dr. Carl Lisher painstakingly reshaped the left ear on October 25, at the same time adding new grafts to the right ear and revising the scars on his left cheek and arm.

Dr. Cannon finalized the shaping of the nose on November 1, 1945. When the bandages finally came away, his face showed signs of improvement but more operations and subsequent rehab operations lay ahead before Heron could finally leave the hospital and think about returning home.

THANKSGIVING DAY fell on November 22 that year. Azelia came to join him for a satisfying dinner and told him of her plans to remain in Phoenixville through his next operation.

On the following Monday, Dr. Lisher refashioned and released Heron's left ear and reworked the mouth deformity.

Ripping his body apart had taken but a fraction of a second in France, while the rebuilding process proved long and arduous. Azelia agonized each time he went under the knife, knowing that the pain level hovered far above what he would admit. Like Heron, she resolved to get past the operations one at a time and did not head home until Saturday for some rest before returning to work the following Monday.

When he first arrived at VFGH, his face looked like that of a statue some maniac worked over with a ball peen hammer. But now, the slow, painstaking efforts of his doctors began to show signs of a payoff. The craggy surfaces once resembling the far side of the moon now looked smooth and he sported a well-shaped nose as well as normal-looking ears.

"MY WORK is finished," Dr. Cannon told him one day. "The army thinks it is time for you to complete the rehab phase so you can be discharged." To Dr. Cannon's trained eye, Heron's face still lacked the proper amount of elasticity. The face appeared waxen and his mouth would remain permanently ajar. His sunken cheeks exaggerated the protrusion of the zygomatic bones, and only a faint delineation marred the roundness of the chin. But there comes a time for tradeoffs between the quality of life and cosmetic improvement. Each subsequent operation represented a short step toward normalcy, but at the price of excessive and debilitating pain.

Photographs of his face taken prior to the first operation compared with his most recent photographs taken a year later showed dramatic improvements. With a pair of sunglasses to conceal the grafted eye sockets, he could move about in public without appearing to the casual observer anyone other than a blind man who suffered physical trauma. Heron could now go out and face the world. The hospital workers who had grown to admire his great courage and lack of self-pity would miss him, his lack of complaints, and the way he always held his head high and walked with pride. He possessed charm, wit and humor, a high level of intelligence, and the courage of a lion. They felt they could do no more for him and at the same time were happy to have played a role in preparing him for the next phase of his life.

"This means you'll be leaving us soon," Dr. Cannon said, almost with regret.

"Getting rid of me, doctor?"

"There's no sense keeping you here any longer. You will find plenty of good doctors at local Veterans hospitals who can handle further repair, many who we trained right here at Valley Forge. You will find good ones at Jamaica Plain and West Roxbury so you won't have to travel far from home to continue your care." He smiled. "Besides, your discharge will be coming through soon. You'll do much better at home with Azelia."

That last statement held the most appeal for the young veteran.

"GOOD MORNING, Larry," a familiar voice said on Monday morning, December 3, 1945.

"Good morning, Father Carroll." Heron immediately recognized the voice of Father Thomas Carroll, a caring priest who dedicated his life not only to providing spiritual guidance to the blind and teaching them to cope with everyday challenges but also how to deal with the psychological effects of blindness, and helping restore feelings of independence and self-worth. Prior to taking over all training for the blind at VFGH, he served as Assistant Director of the Boston Catholic Guild for the Blind, where he became known as a pioneer amongst his peers for remarkable achievements of teaching the use of physical tools and applying the latest research data available to service the needs of the blind.

"Just passing through and thought I'd stop by to congratulate you for making history once again. Not only did you survive the severest of all battlefield injuries and aid our doctors with their development of advanced surgery, you are about to add a third historic achievement to your list."

"And what might that be?"

"You, Sgt. Heron are about to become the first living veteran with a Disabled American Veteran chapter named for him."

"Oh, that."

"Come on. That's never happened before."

"I guess I should be honored, Father. Or perhaps it means I'm really dead and just don't know it yet."

"It's not a trivial matter, I assure you. It's a big deal and we're all very proud of you."

"Thank you Father. I don't mean to sound ungrateful but I can't help thinking about those who gave much more. I often wonder why they were taken and I was left alive."

"God has a purpose."

"I wish I could figure that one out."

"I stopped by for another reason, to give you a heads up on the school we talked about in Avon, Connecticut."

Carroll's role as auxiliary chaplain of the U.S. Army's Experimental Rehabilitation Center at Valley Forge extended to Avon Old Farms Convalescent Hospital, where he divided his time.

"Father, I wonder if you can explain the differences between these army hospitals. Do they all provide the same services as Valley Forge?"

"Letterman General Hospital in San Francisco dates back to the Spanish American War and is the oldest, also the largest until a year ago when they completed Valley Forge. But the injured kept pouring into both hospitals faster than they could keep up, so to alleviate the pressure, construction began last year on Dibble General in Menlo Park."

"How does Old Farms Convalescent Hospital fit in? I mean, what will they do for me that this hospital didn't?"

"Convalescent Hospital is a misnomer because it's really a school rather than a hospital. On January 8, 1944, President Roosevelt signed an executive order declaring that no blinded serviceman would be sent home without proper training. He did so with admirable intentions, but as the numbers of blind servicemen grew, those ready for medical discharge increased exponentially until the president decided none should be released without the benefit of extensive survival training."

President Roosevelt signing a declaration of war with Japan.
Courtesy of the National Archives

"He recognized that veterans' hospitals offered very basic introductory techniques in Braille, typing, writing, and orientation, and this rising tide of blinded veterans could not survive without the benefit of training."

"Just as he needed a quick solution, Old Farms suddenly became available, a ready-made campus perfectly suited for his purposes. The

President signed the order and in just five months, and with little need for funding the vacant prep school became a school for the blind with a specific curriculum to meet specific goals."

"You said it opened just this past June."

"Everything fell into place because of a personal friendship between the President and Mrs. Theodate Pope Riddle."

"Theo…?"

"Born Effie Pope, she changed it to Theodate, which literally means "God's gift. A graduate of Miss Porter's School in Farmington, Connecticut, she set as her schoolgirl dream to one day build an 'indestructible school for boys.' Don't ask what it means or why because I don't know."

"Where did she get the money?"

"She came from money and gained more through marriage, so she built her school in 1926 for eight million dollars and opened it a year later. Then last year a nasty fight with the school's provost over style of management cost the popular provost his job. When he left, the entire faculty resigned in sympathy, forcing the school to close."

"Mrs. Riddle found herself with an empty school just when the President needed it. Old Farms closing was an act of fate that satisfied a convergence of needs, while at the same time accommodating mutual interests."

"What can they teach me at Old Farms that I didn't learn at Valley Forge?"

"The fundamentals you'll need to cope with a lifetime of darkness in a sighted world. The 18-week program features four separate segments: self-care and personal adjustment; physical activities including fitness programs and sports; social recreations like dancing, dinner parties, and movies; and a long list of elective courses such as Braille, typing, manual dexterity, English, music, public speaking, creative writing, as well as courses in industrial and commercial skills."

Fr. Carroll's enthusiasm left Heron feeling like a child on his way to summer camp. Perhaps Avon could restore some of the equilibrium and tranquility that blindness abruptly took away. Fr. Carroll's presence could help facilitate his transition to the school.

At the same time, Heron felt anxious to return home to Azelia. "Do you think it really necessary for me to attend? I get around quite nicely now with a cane."

"Without a doubt, you've made a remarkable recovery and proven most self-reliant. But there's more you can learn, like getting around without a cane. For example, you wouldn't want to be knocking about in your kitchen denting appliances, chipping furniture, or whacking your wife and friends with the end of your cane."

Heron laughed. "Not hardly."

"You already read Braille but when you leave Avon you'll consider yourself a speed-reader by comparison. You'll be better able to care for yourself, play sports, operate appliances, dance, and take in movies. In fact, you'll learn you can do practically anything you set your mind to accomplish."

"I want very much to drive a car again."

Fr. Carroll chewed on that one then laughed aloud, a hearty, congenial laugh. "Okay, we'll teach you but with a few exceptions, like we'll show how to drive a car in neutral gear." The tone of his voice suddenly took on a somber tone. "I understand there's a job already waiting for you."

"At the Draper Corporation as a quality control inspector. Can you imagine a blind person checking quality?"

"You know what? I actually can picture it. And in that case, you'll especially appreciate the final phase at Avon, which includes vocational job training. It stresses the use of hands, sense of touch, manual dexterity, and coordination."

"Sounds good."

"I'm sure you'll do just fine. The program includes working with industrial machinery at one of our local factories. Or, if you prefer, you can practice at a participating store, garage, filling station, or insurance company, whatever you fancy. These tryouts are the capstone of the program. Not only do they improve skills but they also serve as important confidence builders."

Heron came away from the discussion with spirits elevated and awoke next morning with a resolve to expunge the loss of sight from his mind, at least as much as humanly possible. Many people were born blind, never having seen the light of day. At least he enjoyed sight for the first twenty-three years of his life, and fortunately looked upon Azelia for seventeen of them.

FRIDAY, DECEMBER 7, 1945, the anniversary of the bombing of

Pearl Harbor brought hundreds of disabled American veterans to Milford, Massachusetts for an unprecedented ceremony, the dedication of the Lawrence J. Heron DAV Chapter #6, the first and only chapter named for a living veteran.

Heron arrived at the ceremony amid a standing ovation with both Azelia and Emma at his side. Blindness and disfigurement could not mask the raw power of the man who left home a well-conditioned athlete headed for stardom. He departed a local legend and returned a hero who had sacrificed a future for his country. People gathered that night and stood teary-eyed before the tall proud veteran. If he announced his intention to run for Congress at that very moment, he could count on a vote from every person in the room.

Disabled American Veteran Post No. 6 Dedication.
Courtesy of Larry Heron, Jr.

Retiring post commander Peter Frascoti remained standing after everyone took their seats, and when the room grew quiet began to speak. "Thank you ladies and gentlemen for coming to honor this man who by his presence honors each of us. His remarkable rehabilitation against all odds, provided the inspiration that led us to break a time-honored tradition. We secured permission from the Veterans Administration to name this chapter for a living American, marking a first. No DAV chapter in America was ever before dedicated to someone who did not made the supreme sacrifice."

"The pursuits of this fine athlete ended in the span of one fateful second. With no thought of his own safety, he single-handedly assumed the highest risk any person can for the good of his comrades and his country. In so doing, he sacrificed a brilliant future. This man was so badly injured by exploding shells both in his hands and at his feet, that the men in his squad who could not bear to see him suffer drew straws, men who loved him would take his life to end his misery. But he pleaded for them to allow him to return to his lovely wife. Make no mistake she's as much responsible for his survival and rehabilitation as the doctors who worked on him, the power of love kept him alive. He wanted only to return home to the woman who gave him the strength, will and fortitude to defy death."

A murmur passed through the crowd as Azelia blushed and dabbed at her eyes while Emma patted her lightly on the back.

Frascoti turned to face Heron. "Larry, we owe you more than we can ever repay. But for you and men like you, we might not be standing here in a free society as we know it today. God bless you."

Everyone stood for a standing ovation that did not ease until Heron rose to his feet and acknowledged the crowd's approval with nods and a wave of his hand. He felt proud and somewhat embarrassed because he felt like he merely followed his instincts that day, just as anyone in his place would have done.

As the room grew quiet, Frascoti continued. "And now it is my great pleasure to announce that the DAV named Mr. Heron Hero of the Month as part of a national program that honors seriously disabled veterans who demonstrated great courage, ability, and fortitude by rehabilitating themselves. Larry Heron, you astonished the experts while overcoming insurmountable obstacles. You set high standards then surpassed them on the road to becoming self-sustaining. I can't say enough how proud are of you and your accomplishments."

When the applause quieted down, he added, "One final note. We all wish to congratulate you and Mrs. Heron on the expected arrival of your first child."

Yet another standing ovation erupted.

Asked later how he felt about a chapter named for him, Heron answered, "I didn't particularly go for it, but they wanted it so I went along."

He downplayed his heroism, referring to himself as "one of the lucky ones."

A WEEK LATER, Azelia woke with a start, the silence surrounding her seemed palpable and her husband missing from their bed caused immediate alarm. Outside, the scrapping of a snowplow reached a crescendo then slowly died away as she rushed to a front window to peer out at a foot deep accumulation of snow. The branches of tall trees drooped to the ground under heavy loads and from below her window came the sound of a shovel scrapping against the pavement. The few days would bring record accumulations.

Slipping into the robe she left hanging over the back of a chair beside the bed, she descended to the first floor, and as she suspected found Larry outside with a shovel. She wanted to call out for him to come in but instead sat by the window keeping an eye on him as he cleared a path out to the street.

Remarkable, you'd never know he couldn't see a thing.

The path where he shoveled began filling in behind him with drifting snow blown across the yard like desert sand. When she went in her robe to look out the front door, neat mounds of powdery white snow covered the steps leading up to the porch that he previously shoveled, and the mailbox looked like a giant mushroom. Trees beyond where Heron continued shoveling sagged to the breaking point. Yet he continued relentlessly, until he began to feel the weight of each successive load steadily diminish until he knew at last that he had taken an upper hand.

THE FOLLOWING Monday, he returned to VFGH by train for a final checkup and came away with a clean bill of health. Two weeks later, he received his discharge from the hospital and arrived home just in time for the start of the Christmas season, and began to look forward to the start of training at Old Farms in Avon the first week of January. Azelia picked him up at the station, and when she pulled the car into their driveway got out to open the trunk. Heron came around his side and handed her his cane. "You take this, I'll take my bags," he said emphatically.

Azelia knew she should not argue. Let him explore and discover his own limitations. Yet she worried that he might slip and fall so she kept turning to watch his progress as she mounted the front steps. Heron came up the walk feeling obstacles with first one foot then the other until making contact with the first step. When he located it, he mounted the steps toward her then plunked down his suitcases, using them as support in place of his cane.

Larry shoveling a path.
Courtesy of Larry Heron, Jr.

Just then he heard the inner wooden door open and knew someone stood inside, but whoever stood there did not speak a word. His mother would have greeted him, so he waited silently a moment then asked, "Who is it? Azelia, who is at our front door?"

"Why don't we go in and find out," she urged. "Come on." She slipped past him into the house.

Heron followed her inside to a chorus of, "Welcome home!" The cheers came from relatives and close friends, mostly women whose men were away at war.

Later that evening, after the last of the well-wishers departed, Azelia accompanied him to the bedroom to help unpack his things. "What's in this box?" she asked, lifting it from a suitcase. "It's a small box, about the right size of a harmonica."

"Oh that. Better not open it."

"Why not?"

"It's a surprise."

Azelia smiled, her curiosity piqued. "When can I open it?"

"Not until the right moment."

What could the box contain? She could peek now or sneak back and do it later, but that would spoil the surprise. Besides, it seemed important to maintain trust, especially now.

Chapter Nineteen

AVON OLD FARMS
CONVALESCENT SCHOOL

HIS FIRST Christmas at home served only to further Heron's discomfort. Azelia busily hung Christmas decorations he would never see and placed colorfully wrapped gifts under a tree that he could only smell and feel, a sad reminder of happier times past. Azelia tried to understand how he felt and did what she could to make it a joyous time for him, carefully describing each ornament while attaching it to a branch, and when finished, directing his hand to the most meaningful pieces while describing each in great detail.

The enormity of his handicap became ever more apparent when it came time to decide what to buy his wife for Christmas. It filled him with great curiosity as to what she planned to wrap for the new blind man in the house, a package she could tie in newspapers and he'd never know the difference. What should he say when he unwraps it? *Oh what a lovely tie, just my color!* But more importantly, how does a blind man go about buying a gift for his wife and what could he possibly get her, sight unseen?

Azelia's sister, Olga, came to the rescue two days later when she volunteered to take Heron Christmas shopping in downtown Boston. "I go every year at this time. Come with me and I'll help you pick out a nice dress for Azelia. I know that's what she wants."

Filene's Bargain Basement, long a Boston landmark, sat at the junction of Washington and Summer Streets within walking distance of historic sites like the Old State House, the Old Corner Bookstore, and

Faneuil Hall.

In 1946, the area known as Scollay Square included the Government Center, Quincy Market, and parts of the North End, a raucous amalgam of burlesque theaters, vaudeville houses and hot dog stands alongside tattoo parlors, diners, and flophouses.

Further back in time, this marked the site of the lighting of the old North Church by Paul Revere, the house where Thomas Edison designed his first patented invention, and the building where Alexander Graham Bell invented the telephone. An inn once stood at the heart of this location where George Washington actually spent some of his nights.

With parking scarce in Boston's downtown, Olga left the car in a garage near the Common and they walked up Tremont past wrought iron fences and buildings designed by the best known of all American architects, Charles Bulfinch. All along their path, Heron kept a gentle grip on Olga's arm until she slowed at one point. "Let's stop here." She paused then added, "We are standing beside the Granary Burying Grounds."

"I visited this cemetery years ago," Heron told her. "Where Paul Revere, Samuel Adams, and John Hancock were buried."

"That's exactly what's written on this sign. Gives me a chill," she touched his shoulder then told him, "It says that in 1793, on the night of John Hancock's interment, grave robbers removed the signing hand along with other body parts. How awful!"

"Even gives me a chill."

They continued along Bromfield Street to Downtown Crossing, blocks referred to as America's original pedestrian mall, where department store giants Filene's and Jordan Marsh competed on opposite sides of Summer Street. The intersection at Washington and Bromfield teemed with small shops, liquor stores and specialty houses including music stores and joke shops, where practically every product sold in America could be purchased.

Olga searched long and hard but came across nothing she thought Azelia might like in Filene's basement, so they rode a creaky elevator to the third floor controlled by an operator wearing a bellhop uniform that reminded Olga of the little man seen on billboards advertising Philip Morris cigarettes.

After an hour of searching without luck, Olga finally said, "Ah! Here are some nice dresses."

Larry felt the material. "Too coarse."

"Feel this one. It's black and quite sexy. Low back and scalloped front."

"Azelia will not wear anything low in the front."

"Better put it back then, it's a real cleavage number." She took down another. "Now here's a nice dress with three-quarter length sleeves and let's see…it's made of rayon."

"Feels clingy," he said.

"It is."

"What color?"

"Blue. A nice blue, on the dark side of navy blue."

"Do you like it?"

"I'd love it if Fred bought it for me. Rayon is not only real comfortable, it wears and irons well."

"I'll take it."

ON CHRISTMAS EVE, Heron left Azelia sleeping soundly and stole downstairs to place his wrapped present under the tree. By now he could move about the house with relative ease, counting paces in his head. Sitting on the sofa closest to the Christmas tree, he inhaled its familiar fragrance, just as he did years ago when he came down while everyone slept to sit by the family tree in the dark of night contemplating his future.

Another part of the wonderful life he once took for granted, never to experience again. Never a summer sunset or the foliage changing colors in the fall. He could not view his lovely wife or sit behind the wheel of a moving car, play football or observe others playing the game. The same applied to baseball, reading a book, unless written in Braille, or watching a movie. Larry Heron felt like someone stepping off a time machine that traveled back to a time before conveniences like washers and dryers came into existence.

Minutes passed as he sat inhaling the scent of the tree mingled with aromas emanating from a coffee pot percolating in the next room. Lifting his head he formed a new resolve. His eyes and his looks may be gone forever but never his pride and determination. He would not surrender any pleasures in life he could learn to control, like swimming, even if it meant in the ocean with sharks. That's how he planned to live his life, as close to the edge as needed, a life without fear.

If anything, his faith grew stronger, for if God wanted him dead, he surely would not be sitting here now. He could accept that everything happens for a purpose. We may not like what life brings but either we learn to live with it or just give up and die, but not for a minute would he contemplate leaving behind the good things like his wife and their unborn child.

Somehow he knew that when the time came, he would die of natural causes, probably from heart failure like his dad. Meanwhile, why not enjoy everything in his narrow window no matter how small? Clear thinking at times like this put him at ease in a world in which he'd been dealt all the wrong cards, the only cards with which to play the game. He vowed to get on with his life, enjoy it as best he could, and never look back.

HERON ARRIVED by bus at Avon Old Farms Convalescent Hospital on Monday, January 6, 1946 and as he stepped off the bus, staff members came forward anxious to meet the medical marvel who managed to survive shells exploding in his hands, curious as well to check out the great work performed by Dr. Bradford Cannon in rebuilding his mouth, ears, and nose. To those who saw him now, his face appeared somewhat scarred, but the sunglasses softened his appearance and diminished the shock many had experienced when meeting him for the first time.

Avon Old Farms Convalescent School Headmaster's Residence.
Courtesy Avon Old Farms School

During his long bus ride from Philadelphia to Avon, Heron struck up a friendship with the four soldiers accompanying him. Three lost their eyes and the fourth, a man nicknamed "Slim," became injured when he dropped to his knees to dig a foxhole and landed on a mine. Slim could make out dark forms and distinguish red from green traffic lights but his sight continued to deteriorate at a rapid rate and doctors feared he would become totally blind in less than a year.

As Heron disembarked the bus, a guide took his arm and led him along a clear path to a flight of granite steps that rose to a grand foyer echoing their footsteps. He could easily sense the majesty and beauty of their surroundings the moment they entered a large reception hall where Fr. Carroll stood waiting to address the new men. The priest opened by announcing that the school was recently voted the most beautiful campus in America.

In her world travels, Mrs. Theodate Pope Riddle fell in love with the Tudor style architecture in the Cotswold Hills, a section west of London referred to as 'the heart of England.' She decided the Tudor-Cotswold style perfect for the 3000-acre dream-school she envisioned, and started in with its design and construction immediately upon her return.

Theodate Pope Riddle (1867-1946).
Avon Old Farms School owner, designer and founder.
Courtesy of Archives, Hill-Stead Museum, Farmington, CT

When completed, the campus resembled the quaint architectural style of centuries-old English schools like Eton and Oxford, with massive walls of brownstone cut from local quarries. The doors, woodwork, and sturdy beams throughout were hand-hewn and originated from the massive trees that grew in abundance on her property. The red slate rooftops and hand-wrought ironwork featured within and outside the school added color, beauty and depth to pageantry.

The irony of blind trainees touring the most beautiful school in the land was not lost on Heron, who could only imagine such splendor. Upon completion of his admission, an assigned guide accompanied him to his room. As the two men entered the building that housed his dormitory, the guide explained, "You will find a sliding wooden latch on each door." He spoke as he led Heron to the steps to the room, "Stop here. About a foot in front of you are stone steps curving up clockwise. You will find every dorm has the same set of stairs, all curving to the right like those found in European castles."

"Why is that, curving right I mean?"

"Like now, most people in medieval times were right-handed and needed more open space on the right side to wield their swords. Stairwells that rose clockwise left attackers coming up the stairs with their sword hands restricted by the curve of the wall, making it difficult to swing. Defenders backing up the stairs could hug the inside wall, leaving their sword hands more room to swing. Similarly, medieval swordplay led to driving cars on the left in England, so that riders on horseback or in a carriage could swing their swords in the open space on their right without fear of catching it on walls, branches or corners of buildings."

They reached the second landing and walked another twenty paces before the guide again asked Heron to stop. "The door to your room is directly in front of you, the doorknob to your right. Go ahead and let yourself in."

When they stood inside, the guide described the room as fairly small but efficiently laid-out. Next, he introduced Heron to a wood frame bed in one corner, a table and chair below a window sill, and a phone that sat on a small desk. The desk provided a flat surface to lay down a book written in braille or items like a comb from his pocket. Nice, Heron thought. He might even try his hand at writing a letter at the desk, for he had learned through practice to keep his words in a straight line, and that by marking the line with a finger from his left hand to gage how much to drop down, he could start writing on the next line.

"Your suitcase is just right of the door as you exit the room," the attendant informed him. After a quick tour of the attached bathroom, he led Heron to a wardrobe closet and stood by while he unpacked his own shirts, pants, and jackets, draping them over hangers then lining them up side-by-side in his closet. When finished, the attendant complimented Heron. "They look just fine. Sure you can't see?"

Heron could not stifle a proud smile.

"You can finish unpacking the rest of your suitcase and put things in this drawer that pulls out from under your bed, in whatever order works best for you. That way, you'll be able to find things later when you need them."

When Heron finished unpacking his suitcase and stored everything carefully away, the guide said, "Okay, now lose the cane, Mr. Heron. You won't need it again until the day you leave campus."

"But I depend on it for..."

"We're going to teach you how to get along without a cane. Especially indoors."

"I'm expected to go outdoors without one?"

"You'll use it only in town. No canes allowed on campus. We call it 'mobility training.' Canes become unwieldy inside homes or offices where it's too easy to knock something over, damage furniture, or whack someone on the shins."

"Mobility means no cane."

"You've heard the expression blind as a bat?" It was more a statement than a question. "Mobility emphasizes the reflected sound approach. In theory, when a blind or blindfolded person approaches an obstruction, like a large tree, car, or building for example, he or she will usually sense the object before crashing into it."

"I hope you're right about that."

"I'm sure you've felt it from time to time."

"Do all blind people have this...sixth sense?"

"Not to the same degree, no. That's why I said 'in theory.' Very few have the confidence or makeup to fully develop it. Most people never get the hang of it."

"I see."

The attendant caught his play on words and they both laughed.

That night Heron joined the line queued in front of the mess hall, which to a sighted person appeared like a faithful reproduction of an old English baronial hall. From the echoes of footsteps and the general feel of the room, he could make sense of its enormous size, confirmed minutes

later when their guide mentioned it could accommodate all post personnel at one time. The rear entry behind the table where he took a seat led to
another large hall where weekly dances took place.

Most trainees introduced themselves as young and physically fit
who just couldn't see, some had lost fingers or perhaps a limb, while others lost both arms or both legs confining them to wheel chairs. Though a
handful went through plastic surgery, none faced as many operations as
Heron.

The following morning after breakfast, the newcomers gathered
in a small conference room to listen to an address by the school's commanding officer, ophthalmologist Col. Fredrick H. Thorne. Like the other
patients, Heron did not know the color of the Colonel's eyes, whether or
nor he wore glasses or sported a mustache or beard, but he did take note of
his embullient and outgoing manner.

"Good morning, men," he began. "Welcome to Old Farms Convalescent Hospital, which occupies a unique position in the Army's program
of caring for wounded servicemen. You will find the school's sole function
is to prepare you for readjustment to civilian life."

"First a bit of history. In 1943, our hospital staffs in Phoenixville
and San Francisco, overwhelmed by the flood of blinded patients physically ready for medical discharge, found that they lacked the proper rehabilitation and training to prepare patients coping with the permanent loss
of sight. Based on their cries for help, the Secretary of War, the Secretary
of Navy, the War Commission Chairman, as well as the Veterans Administration Administrator…" He paused to allow for some chuckles to subside
"Yes, quite a mouthful. They all first agreed Army and Navy blinded personnel fall under the responsibility of the Army Medical Department. The
answer they came up with was to establish the Avon Old Farms Convalescent Hospital."

"You will find the school welcomes you with an understanding of
your needs and the problems you face as individuals. Our goal is to offer
many courses to choose from, and those of you who work diligently will
find yourselves amply repaid in the years ahead."

"When your stay with us has ended, each of you will leave here
with not only the proper training to help you find a job and earn a living,
but also with confidence in your own ability to live a useful life in your
community."

HERON TOOK a special tour of the facilities with a group of fellow students. On this balmy day, the smell of horses greeted him long before he heard their whinnies. As at all their stops that day, the guides halted the blinded audience just outside the stables to offer a description of immediate surroundings and to acclimate the men to their overall environment. "The stable in front of you was built in the style of an old Norman manor house of stone and half-timber with a long sloping roof."

"Horseback riding is just one of many recreations you will find. We will introduce you to facilities on campus where you can bowl, swim, roller skate, fish and play golf. If you think blindness will prevent you from participating in these activities, think again. In the summer months, we will even show you how to go boating on the Farmington River, if you desire."

Just beyond the stables the group halted again, this time before a massive circular tower made of red brick. "This water tower is a famous landmark that rises majestically above the campus with its signature tall silo and cone-shaped roof that blends nicely with the rest of Old Farms."

Upon returning to the mess hall for lunch and to warm their freezing toes, the tour recommenced at the motor pool. "It too looks like something you might find in the English countryside but inside you will find ambulances, buses, staff cars, and a fleet of trucks, which those of you who want to become mechanics will learn to maintain." This latter comment elicited sounds of approval from the mechanically inclined.

At dinner that night and throughout the first week, Heron picked up interesting statistics regarding fellow trainees, such as the range in age ran from 18 to 45, in education from fourth grade to PhD, and in rank from private to colonel. Except for blindness, most men left the battlefield in fairly good physical condition. Heron received the most severe injuries and spent more time in the hospital before coming here, but at least he still had limbs and hearing.

Each day began with a half-hour session allowing the veterans to air complaints about food, facilities, transportation, or any subject they cared to discuss. A visit and talk by the CEO from a company like Champion Spark Plug, Royal Typewriter, or Fuller Brush would follow, and the speaker would attempt to entice trainees to consider an interview with their respective companies regarding future job possibilities.

Plenty of social activities took place weekly. The men could visit town bars and the school held weekly dances where many came in contact with women destined to become future wives.

WHEN AZELIA visited Heron the following weekend, she could not believe her eyes. "How did you learn to get around without your cane?"

"Practice," he said with a smile. "I count paces and memorize direction, drawing a mental map for each destination."

"You were always independent but I can't believe you don't need your cane."

"The human mind is survival driven and retains information that allows me to hone in where I want to go. Also, I get a feel for what's ahead, pick up subtle sounds and feelings. It's hard to explain to someone with sight."

"You've always had a remarkable memory."

"I walked these grounds every day mapping them out, but not without mishap, as noted by the bruises on my face?"

"Been meaning to ask how you got them."

"I ran into Big Bertha. Come, I'll show you."

He led her along a path at a remarkably brisk pace to a sturdy elm of considerable girth rooted next to the main walkway by the school's entrance. "I practiced walking around campus on my own and moved pretty fast until wham, I smacked into what I felt like a brick wall." He reached out to place a hand on the tree. "At least it felt like one. Anyway, I didn't know what hit me until I reached out and touched it." He touched the bark of the tree as he spoke. "Then I felt blood trickling down my new nose and thought 'just what I need,' another scar."

"One of the attendants came by just then and laughed. 'Sounds like you just met Bertha," he said. 'She's got plenty of skin, hair, and blood on her. Sooner or later, every trainee makes her acquaintance.' He spoke almost affectionately of the damned thing."

Azelia grimaced. "Your poor nose."

"It was my first day outside without a cane. I assure you it will never happen again."

Azelia took in the beauty and majesty of the school as they moved past the ski hill, bowling alley, and golf course. "Oh, look. Horses!" she exclaimed.

"I wish I could," Heron chuckled, as he listened to the sound of hooves clopping on hard ground.

They stopped while an attendant led three trainees past them on horseback along a bridle path plowed several yards wide. Azelia felt comfortable around horses, having ridden them often at a friend's stable in Mendon. "The horses know their way around," Heron told her. "Most trainees prefer to wait until the weather improves before trying to ride them."

"Maybe we can do some riding together next summer," Azelia said.

"A month from now I'll be home," he said encouragingly. "I can hardly wait."

"Neither can I," she said, giving him a hug.

The air smelled like clean laundry pulled off a clothesline on a crisp winter's day. The sun's rays felt warm and soothing as Azelia observed two men on a bicycle-built-for-two peddling up the street, the sighted rider guiding a blinded man. She watched blind men working on engines in the motor pool and others practicing on instruments in the music room. At the carpentry shop she observed four trainees building cabinets while three others operated the school's switchboard.

Had she not seen these men participating in so many diverse activities with her own eyes, she never would believe it possible. It raised her hopes that in a month, Larry would leave here not committed to a life of mental and physical torpor, that he could instead look forward to a near normal future existence.

WITH EYES focused on the moon as it traverses her window, she forms a mental picture of her "Adonis," only this time she sees him with his eyes intact, the way he looked when he left for war. In time she learned that when laying beside him, if she closed her eyes she could visualize how he once looked, as he appeared when the girls from St. Mary's commented that he looked like the Greek god Adonis, and thought him the best-looking man in town.

She recalls how he looked when he made love to her face-to-face, before the injuries. Now she could keep her eyes open and look at him, superimposing the missing blue eyes in her mind, covering over his new eye-less face, seeing him just as she once remembered. One advantage, no longer must she worry that another woman might try to steal him, something she readily accepted. In the end, Azelia slept with Adonis and made love with him, content that no one else would have him. His body, his mind, his spirit belonged to her and she never wavered in her love for him.

When he went away for any length of time, to the school for example, she missed him but felt safe with him in Avon learning to cope with blindness, and not subjected to the pain of more operations.

Azelia with Larry reclining at Avon Old Farms School.
Courtesy of Larry Heron

HERON TOOK to training with ease. While many struggled with advanced Braille, he mastered it quickly, earning high grades in both Braille and typing. He spent a great deal of time working with leather and could soon craft beautiful wallets and women's pocketbooks. He also worked on *The Quadrangle Review*, a newspaper published exclusively by trainees.

But he most enjoyed working to improve his singing and growing his repertoire. Encouragement came from other trainees, and then one day Col. Thorne asked if he would mind singing a few songs at the next weekly post dance. It caught on immediately and after he continued entertaining at the dance each week, Fr. Carroll's invited him to solo at one of church services. Col. Thorne met with Heron after the services to discuss his future. "You should consider singing professionally. Many celebrities who came to entertain us would be more than happy to help you get started."

Heron thanked him, but politely turned down the offer because it he felt sure it would provide an uncertain future, and any success might lead to travel and associated hardships. "I'm just looking forward to returning home with Azelia to raise a family."

T w o m o n t h s later Azelia's brother, Guido Noferi, returned home after serving in the Navy throughout much of the war. Then on March 15, he and his good friend, Ed Kalpagian drove Azelia to Avon to pick up Heron to drive him home in time for the new baby to arrive. Once Azelia returned home safely with the new baby, they would transport Heron to Old Farms so he could complete his training.

Guido and Ed, like so many patriotic citizens of the greatest generation, lied about their ages to join the Navy, and both saw action during the war. Immediately after they joined the Navy reassigned Ed to the Marines where he served as a medic throughout the war, serving in horrific battles fought against the Japanese in Iwo Jima and Tarawa.

Fr. Carroll quickly spotted their party entering the reception area, mostly because Azelia stood out as the most attractive woman in the room, and in part because she also appeared noticeably pregnant.

When he greeted them, Azelia read strength in the depth of eyes as gray as wood-smoke on a cold winter's day. A force emanated from them that she found uplifting even before the priest had uttered a word. Following introductions, he told her, "Larry will be joining us shortly. An attendant will locate him then bring down his suitcase."

Seated in the library at that precise moment, Heron lowered a book written in Braille and thought how much the school served to broaden his outlook. He now knew the true meaning of "mobility," and how to maximize the use of the senses explosions did not destroy, his touch and hearing. Fr. Carroll found truly amazing his "sixth sense" and uncanny ability to recognize people by their voices, even people he hadn't spoken to since joining the Army, and he could identify others by the sound of their walk and the shuffle of their feet.

Many of the blind trainees could properly interpret the subdued bustle of changing shifts taking place when a new wave of orderlies, nurses, technicians, administrators, and doctors arrived to take the next shift, and the sound of the last shift exchanging goodnights and departing. But only the most discerning could identify the next set of footsteps as those of an orderly named Jim Hanson.

"Hi, Jim," Heron said even before the orderly spotted him. The impressed orderly answered, "Oh. There you are, Larry. Your wife's waiting in the reception area."

As Heron's visitors sat waiting, Fr. Carroll asked Azelia, "Did you know we offered Larry a Seeing Eye dog?"

"Yes," Azelia answered. "And he turned it down. Do you think it was a mistake?"

"To be honest, I believe a seeing-eye dog would only slow him down. He gets around exceptionally well without help. I don't know how he does it, but I've watched him walk around town as though he has eyes."

"He counts paces and forms mental pictures in his mind, his photographic mind."

Ah, yes," the priest said. "And here he comes now. Look at him. You'd swear he can see right where he's going."

AZELIA GAVE birth to Patricia Ann Heron on St. Patrick's Day, Sunday, March 17, 1946. Heron could only "see" little Patty with his fingertips but when he finished counting the tiny fingers and toes, he exclaimed, "She's absolutely gorgeous."

Patty brought meaning back into his life with renewed hope for a future. Before she came along, he merely existed, now he felt real purpose, a reason to work harder and set new goals. Winter may follow fall in terms of the past, but the birth of Patty proved that sunshine also follows rain. *Now I have a daughter, la raison d'être.* Sunshine returned to his life.

Azelia watched as he held the baby tenderly, but agonized when she considered the suffering weighing heavy on his heart. Though his daughter gave him delight, he would never see her smile, know the actual color of her hair, or watch her pass through the various stages of growth to adulthood. Azelia silently cursed the war for that, and Hitler for his maniacal pursuits, praying there would never be another war.

HERON RETURNED to Old Farms filled with renewed hope and immediately became involved in the formation of the Blinded Veterans Association, a historically significant organization that received its charter from Congress in 1958, with intent to continue promoting the welfare of blinded veterans.

On Friday, Heron accepted an invitation to spend a weekend at the rectory with Fr. Carroll as part of his rehabilitation program. For dinner on Friday night, the chaplain prepared haddock dipped in egg whites and coated with breadcrumbs, cornmeal, and lemon-pepper seasoning, served to his guest along with skillet-fried potatoes.

After dinner, the two men sat by a fire listening to the periodic crackle of logs cut from apple trees that emitted a bewitching fragrance.

Fr. Carroll poured two snifters of Janneau Grand Armagnac, extra old with a bouquet that could smooth the hackles of an angry bull. They lifted their glasses and Heron toasted, "Here's looking at you, Father." It took a moment to sink in then both laughed at Heron's choice of words, as he quickly added, "Don't I wish."

Fr. Carroll studied Heron a moment then stated, "One thing I've noticed about you, Larry." He paused. "I admire the strength you demonstrate by accepting your fate without bitterness or agitation. Many soldiers who suffered blindness admitted they gave up believing in our Lord but you show no bitterness."

Heron used his free hand to locate a bare spot on the coffee table to lower his glass then leaned back against the soft cushions of his wing chair. "I don't know. There are times when I listen to football or baseball that I find it difficult to keep up my spirits. I miss running to catch a pass, drive a car, things like that. And at times I find myself weakening as I wonder, why me? That's when I start questioning His existence."

"There's not much I can offer in the way of guidance with respect to your tragic fate but Emerson wrote about greatness born of suffering. Being shoved around, bullied, even defeated, humbles us and teaches us the real facts of life. We all have doubts from time to time; that's quite normal. But you seem to handle blindness better than most, and appear surer of yourself than many with the ability to see."

"I try not to look back too often or stew about what I can't control."

Fr. Carroll leaned his elbows on his knees and tented his hands. "You know, I've worked with blind people for most of my life and find you quite unique. You have a quiet and singular confidence, perhaps a holdover from your earlier life. You moved well past the trauma associated with sudden blindness, much sooner than I would have expected."

"Believe me, I'm not always as sure as I may appear."

"Then you hide it well. I wonder if you'd mind answering a few questions. You see, I'm writing a book about my experiences working with the blind. These dinners with patients give me the opportunity to ask questions one-on-one, and your answers could prove helpful."

"I'm happy to try."

"I think you've already answered my first question which concerns self-image. You're not like most. The trend for most people suddenly losing their sight is a significant loss of self-image."

"I haven't experienced much of that, perhaps because I am married to someone very special who does not allow it."

"You give her much of the credit and that is commendable but I think there's a lot more to it. You possess something that helps you cope, the confidence you developed playing sports, the knowledge that you can succeed at anything you try."

"I suppose."

"Do you go along with the perception that people develop extra power in their other senses to compensate for loss of sight? Much has been written about it but mostly by sighted writers. Do you feel your other senses have strengthened and improved, and if so, to what degree?"

"I don't know that they've grown stronger or simply that I rely more heavily on them since losing my vision. My hearing was always exceptional as well as my ability to distinguish people by their voices. Now that my life depends on it, I tend to lean more heavily on my remaining senses, so much so, they may have improved."

Fr. Carroll nodded as though he understood then realized Heron couldn't see the movement. "Understanding such feelings and addressing them is the key to learning how to live with blindness. I hope that when it's finished, my book will prove helpful to others."

ON SATURDAY afternoon, while drinking beer and talking for hours with his host, Heron discovered Carroll had studied Greek, Latin, and philosophy at Holy Cross. Graduating in 1932 from St. John's University, he was ordained in 1938, and began working with blinded veterans at the start of the war.

Like an athlete who practices hours on end to develop his skill and hone his trade, Fr. Carroll devoted long hours of his life to helping the blind as his passion. In time, the publication of his techniques and documentation of his results led to national recognition for finding ways to help the blind, and it aided other practitioners who sought to apply his tested methods at schools around the country and the world. Fr. Carroll not only founded St. Paul's Rehabilitation Center for the Blind, he also developed St. Raphael's Geriatric Adjustment Center and the American Center for Research in Blindness and Rehabilitation. Subsequently, he was appointed the National Chaplain of the Blinded Veterans Association, serving as a member of the President's Council Committee on the Employment of the Physically Handicapped.

Heron found the learned and innovative priest kind and thoughtful, and quite skillful when they sat down to play cards. After dinner, the priest introduced him to the new long white cane introduced recently by Dr. Richard Hoover, a cane that would one day become the standard traveling tool employed by blind people nationwide.

When it came time to return to his dormitory, Fr. Carroll presented Heron with the cane as a keepsake. "Something to remember me by," he said with a friendly chuckle.

Heron took the cane from him, tapped the floor a few times then reached out to shake his hand. "Thank you, Father. It will always hold a special meaning. I'll put it away to take home with me."

Fr. Carroll, like Fr. Connors, proved a rare priest truly dedicated to his work, strong in his spiritual beliefs, and high on fortitude. Both priests played a strong hand in helping Heron hold on to his faith.

On the following Friday, his last day at Avon Old Farms, Heron nerves felt raw. The next morning Azelia would arrive to take him home and surprisingly, he did not feel anxious to leave this fine institution. In the span of a few weeks, Avon Old Farms had become a sort of home, surrounding him with hundreds of men sharing the same afflictions. Each man grappled like he did for some semblance of normalcy, each capable of taking the same missteps in life. Each in turn bumped a tree or tripped over misplaced objects, just as he did, but all experienced the same feelings. Mutual experiences and understanding of what it took to survive in a dark world separated this special band of brothers from sighted people who could never imagine the turmoil raging inside.

Late that night, he sat on a bench alone under a sky peppered with stars he would never see, listened to the hollow barking of a dog in the distance, and worried about tomorrow. The world waiting outside these walls represented an unforgiving, even hostile environment. At that moment, opposing forces struggled within him: fear that leaving the shelter of Old Farms behind would render him disconsolate and off balance; and anticipation of happiness resulting from his return to the woman he loved to reestablish their lives together in Hopedale.

Heron did not face such fears alone. By the time the Army's occupation of the school would end in 1947, eight hundred-fifty trainees would have passed through its portals, with every last man sharing similar feelings of sadness and fear. None of the blind men looked forward to leaving the comfort and companionship of Avon Old Farms Convalescent Hospital

208

to reenter a world filled with uncertainty.

Azelia in 1946 at age 27. *Courtesy of Olga Kalpagian.*

THEY WAITED in the warmth of his kitchen on Wednesday, May 1, 1946, his family, friends, and two-month-old daughter, Patty. "Surprise!" Patty did not see a blind man or a disfigured man come through the door, she saw "Daddy!" Well, she didn't exactly call it out, she made noises like "Do dah," and too young to rush forward to throw herself into his open arms, lay waving them excitedly about her like windmills.

It felt good to inhale the lusty smells of home cooking and hear voices filled with laughter and good cheer that put him very much at ease. During his stay at Old Farms, Heron made some important adjustments and formed some positive new ideas. For one thing, he no longer felt the need to apologize for his ugly appearance, and he stood prepared to tackle just about anything that interested him, without the need to solicit help from anyone.

After the initial shock of seeing him for the first time, people quickly rose above their feelings with the quick realization that he possessed the same level of guts and determination he demonstrated on fields of friendly strife before leaving to join the war. Larry Heron remained fiercely independent and as affable, sociable, and outgoing as ever.

One of the first places Heron wanted to revisit upon his return was a restaurant called Milly Mitchell's located on Nipmuc Pond. For months on end, he missed and now craved the, heavenly New England-style fried clams. On their first visit, Milly stopped at his table to shake Heron's hand. "Welcome home, Larry. This one's on me." She gave Azelia a hug, saying, "I remember when you used to come here with your dad - such a nice man."

Heron asked, "Did you know that once you leave New England, it's like fried clams don't exist? The ones with the bellies, that is."

"Yeah, I know."

"All the time I was away, I had this craving. Nowhere, not even on the Cape, can you find fried clams that taste this good."

"Well thank you. Larry. That's real nice."

Their final stop before heading home was at Lowell's Dairy for a double scoop of homemade ice cream, chocolate atop a scoop of vanilla. The rich dark chocolate had a unique nutty cocoa flavor, and the second-to-none creamy vanilla not only satisfied, but definitely became addictive.

Heron enjoyed revisiting old haunts where past friends came forward to welcome the returning hero. Rumors preceding his return claimed that Heron threw himself on a grenade to save his comrades. Not everyone knew the real story and he did not care to elaborate because reliving the event invited much trauma. "The day started out awful and went downhill after that," he would tell them.

CHAPTER TWENTY

THE ABSENCE OF LIGHT

ON A FRIDAY evening in mid-summer, Heron came down to the kitchen carrying the mysterious box Azelia unpacked from his suitcase then completely let slip from her mind. Now Heron held it in his hand, piquing her curiosity once again. "So what is in that box? You promised to tell but kept me in suspense all this time."

"Open it," Heron said, handing it to her.

Azelia slowly lifted the cover to peek inside and felt two hard objects wrapped in tissue. She smiled and looked up at Heron as she tugged at the tissue and out they popped. The box slipping from her grasp, and when it hit the floor, twin objects rolled out like two huge oblong marbles that she recognized as a pair of eyeballs alternately looking up as they rolled past her, as though searching the room. "Dear God!" she gasped,

"What's going on?" Heron asked.

"I dropped the box and these... things rolled out that look like eyes." He could only imagine how eerie they must appear.

Looking down at them now, she found their colors flat and lifeless, totally lacking in depth, and monstrously huge outside their intended sockets.

"Larry, they're..." She couldn't find the words.

"Samples," he finished. "Should I decide to go with fake eyeballs, they'll custom fit new ones to each socket using a gel..."

"And they'll look like these?"

"Yes. They provided these samples so you could help me decide."

She gazed down at the eye balls staring vacantly in different directions. "Honey, I've already decided. I prefer you just as you are."

"I thought if you liked them..."

She thought of more operations he must endure to remove the skin Dr. Vance Bradford grafted over the sockets. "Your sunglasses work just fine. Trust me. These things… they just don't look real."

"Okay. If that's how you feel, and don't worry I'm not upset. It'll be a pleasure to skip the trouble of having them fitted."

She looked long and hard at him to determine he wasn't just saying it to please her then lifted the eyeballs one at a time as though lifting dead rats by their tails, placing them side-by-side in the box. On her next trip upstairs, she would load them into a shoebox and place the box at the rear of their closet.

The "rock" Heron leaned on. *Courtesy of Larry Heron, Jr.*

ON SATURDAY morning, Heron pulled on a clean white shirt and tucked it into a pair of jeans still carrying the scent of fresh air from outdoor drying then climbed into the car so Azelia could drive the family to Amelia's house on Dutcher Street for a visit that included a lazy afternoon cookout. That afternoon everyone sat around rehashing the news and weather while sipping from beer mugs or glasses of wine. Two hours had passed when Azelia asked Heron if he would like to take a ride to the Rustic Bridge, a beautiful setting less than a mile from the house, where they could stretch and spend some quiet time alone.

Amelia, who overheard said, "Why not right now? Go. Go. You'll never get a better chance. Patty is happy playing with Linda and won't even miss you." Close in age, their daughters connected like sisters. "When they get bored, I'll take them next door to visit Mrs. Kalpagian, who always likes the kids to visit."

The temperatures that day hovered around 75 degrees with overhead clouds drifting like sheer lace curtains across a cerulean sky, breaking up lazily beneath the warm sunshine. After leaving their parked car to approach the Mill River swirling toward Hopedale Pond, Azelia described the tall clumps of marsh rising from the fen on either side of the causeway, their reeds bowing in the slight breeze as if inviting the pair to cross the bridge.

It felt good breathing the clean air while listening to whispering wind sift through the pines. Nothing served better to ease tensions gathering inside like threatening storm clouds than the warmth of sunshine on such an afternoon. Upon reaching the two stonewalls running along both edges of the bridge, each found smooth areas along the wall facing the side with the widest expanse of water.

Heron asked, "Tell me what you see."

"It's like something Thoreau would describe and Rockwell would love to paint, clear sky, fish jumping, and two people fishing in a small boat on the other side of the pond."

The scent of pine lingered and as serene minutes slid past, Heron listened to the sounds of birds and the wind now rustling through trees.

"It's so peaceful," he said at last. "Like that time at Old Farms when I went ice fishing."

"I don't know if I'd like that. Too cold."

"Sometimes, but the cold never bothered me and they had to drag me inside to get me off the pond."

"You are pretty amazing," she said. "Nothing slows you down and nothing seems impossible."

"What about you?"

"Me?"

"I used to lie in the hospital thinking what I would do if you'd left me," he said, wistfully. "Even now I can't believe you stuck through it all. You such perfection and me...damaged goods."

"Don't be silly,' Azelia said. "The only thing I ever wanted was to be with you. I was miserable the whole time you were away and I will go on loving you, grow old with you, and die with you. Selfishly, I hope I go first, for I could never survive without you."

He hesitated a moment. "You can't know how much that means to me. Everything's been taken from me except the most important gift of all, you, the one constant in my life, the only thing in this crazy world that makes any sense. With you in my life, I can accept what's happened and live with it. Without you – well that would be unthinkable."

"That should never give you cause for concern."

THE FOLLOWING Tuesday, Ed drove Heron and Guido to Dave's Barbershop on Purchase Street in Milford and as with sound, Heron could attune to the smells of life cooking odors, cigarette smoke, exhaust fumes, flowers, ozone, not to exclude awful smells like garbage and decaying fish. Even before they turned off Main Street onto Pine, Heron picked up a whiff of the barbershop a half-block away, unmistakable combinations of hair tonic, shaving lather, cigar smoke, and human musk.

Inside, Guido and Heron took seats with a view toward the storefront where a variety of tropical plants sat on display in two large windows, one on either side of the shop's entrance. Overhead, three ceiling fans gently stirred the air while emitting a hum and a repetitious clacking as soothing as the motion of a moving train, guaranteed to set patron's heads nodding within minutes of sitting down to wait their turn.

Ed stood looked down the street through a window where he spotted Charlie turning off Main Street a half block away, headed toward the shop. "Here comes Charlie Everett," he mischievously whispered to Heron. "He's wearing a Red Sox baseball cap, a blue sports jacket and carrying a large brown paper bag." Before the unsuspecting target could spot him, Ed ducked into the storage room while Guido buried his face in a newspaper. If Charlie spotted them together, he might suspect the two clowns cooked something up.

A bell jangled when the door opened to admit Charlie Everett to the barbershop. Heron spoke up as Charlie sat down to wait his turn.

"Charlie! You sure are a sight for sore eyes!"

Charlie turned to stare at Heron in wonderment.

Dave said, "Oh. Hi Charlie," and continued cutting hair, as the confused patron glanced around at other customers to see if they heard what Heron said. No one paid the slightest attention, and instead sat reading magazines or dozing.

"What? How?" Charlie scratched his head.

Dave stopped snipping and cocked a discerning eye.

Heron said, "Nice jacket, Charlie. Blue looks good on you. Is it new?"

Everyone seated in the shop fought to stifle laughs while Charlie sat convinced that Heron could see him through the sunglasses. But that lasted until common sense kicked in and an embarrassed smirk crossed his face. "Hey, what's going on here?"

Everyone kept on reading magazines or sat with eyes closed or simply stared straight ahead, all the while fighting to keep from laughing.

Heron continued the ruse. "Why don't you put that bag down and take a load off?"

"Okay, you heard the bag crinkling. That's it. Nice try, Larry."

"You look good in that Red Sox baseball cap. It looks like the real deal."

Charlie grinned, sheepishly. "Come on. You can't see." He lost the smile. "Can you? Nah. You're just putting me on, aren't you? But how'd you know...?"

Laughter came from behind the storeroom door sitting slightly ajar, drawing Charlie's attention. He peeked through the crack and saw movement but before taking a look inside saw the newspaper in Guido held shudder in his hands as he tried unsuccessfully to suppress a laugh.

"Okay you guys!" Charlie pushed open the storeroom door and saw Ed bent over with laughter. Then Charlie turned around and ripped the newspaper from Guido's hands. "You bastards," he chided. "Having fun?"

The barbershop erupted in spontaneous laughter.

MAKING LIGHT of his blindness helped Heron to deal with it. When someone admired his necktie, he would say, "Glad you like it. Picked it out myself." Or when he'd drop by the soda fountain at the Hopedale Drug Store, he might ask the friendly waitress, "Joanie, what did you do to your hair? I liked it better the other way." Jokes and unexpected comments made for warm relationships and directed attention away from his afflictions.

On daily walks about town, someone invariably would call out, "Hey Larry, how's it going?" Unless it came from a total stranger, he'd fire back, "Just fine, and how about you, Billy?" Then Billy, or Jimmy, or whoever would drop a jaw in wonderment, asking, "How did you know it was me?"

THE INVITATION arrived for him to march in the July 4th parade that typically started out at the American Legion Post in the town center and ended at the Hopedale cemetery. American flags mounted on ten-foot poles planted about thirty feet apart along Hopedale Street complemented red, white, and blue banners draped in upper corners of the town hall. The clap of hooves echoing off the pavement announced the arrival of caissons towed behind black horses. On windy days, flags would flutter to life beside wreathes or flap against headstones in the town cemetery, and weather permitting at the start, low-flying National Guard fighter jets would make low passes and tilt their wings before heading off to other towns up and down the South Shore.

On the morning that marked Heron's participation in his first July 4th parade, Azelia delivered him to a designated spot where members of the American Legion and Disabled Veterans began forming for the parade. He wore his best tan suit with a white shirt, maroon tie and a navy blue American Legion cap astride his head. "You needn't worry," one of the marchers told her. "We'll keep our eyes out for him."

With some reluctance, she joined members of her family staked out at a spot across from the pharmacy and felt the excitement as throngs lined both sides of the street and the band started playing *Stars and Stripes Forever.* Patriotic cheers emanated from the crowds and Azelia felt a surge of nervous anticipation as she caught sight of Larry marching up the street towards her.

As the band grew louder, she maneuvered herself closer to the edge of the sidewalk for a better view.

"Well I'll be!" said Charlie Everett, standing on her right. "Look at Larry! You'd never know he can't see a thing!"

Azelia blushed with pride. He hadn't told her he would actually *lead* the parade as Grand Marshal. She admitted to herself that he cut a neat figure with head high and cane pumping in his right hand like a baton. His left hand scarcely touched the sleeve of the officer marching beside him as he moved forward in step with the others. Behind them marched a color guard made up of members of the Armed Forces bearing the American flag at the center, flanked by American Legion and VFW post flags and on each end flag bearers, one soldier and one sailor, each shouldering a rifle. Most townsfolk lining the parade route took pictures and leaned over to catch a glimpse of their real live heroes. The crowd seemed as thrilled as Azelia at a stirring sight none could ever forget.

AS WORD OF Heron's sweet tenor voice spread, the Sacred Heart Catholic Church on Hopedale Street invited him to join their choir and at Midnight Mass on his first Christmas Eve at home. People showed up for mass from neighboring towns to hear him sing Ave Maria. Surrounded by colorful poinsettias and the scent of candles and firs, they listened to his voice resonate throughout the hollow church like an echo chamber. When finished, one could hear a pin drop, and enough tears spilled that night to fill an Olympic-sized pool.

Barely a month later, Heron received an invitation to join the Milford Area Singers and began performing at hospitals, nursing facilities, weddings, and important events throughout the Blackstone Valley. Word of Heron's heroics preceded him and soon the well-known figure received invitations to appear as honored guest to deliver patriotic speeches at various functions throughout New England. Meanwhile, he would never miss a day of work at the Draper Corporation, except to return to the veterans hospital for more operations.

Music held special meaning in the Heron household, a place where friends and relatives congregated on holidays. At those special times, Azelia's sister, Olga, would alternate with Ed Kalpagian at the piano while others sang along with Heron.

When a reporter from the Worcester Gazette interviewed Heron for an article about his singing, he also asked questions pertaining to blindness. The quote he printed read: *It never pays to look back with regret, because you can't change your past, it's best to stay in the present. There are things that happen that I have no power to change, those I leave to God while focusing my energies on what is in my power to change.*

Every few months he would return to the West Roxbury Veterans Hospital where doctors continued work on the reconstruction of his face. Perhaps as a subconscious effort to compensate for disfigurement, he began paying closer attention to his appearance, something he could his control, and soon developed an almost fanatical desire to properly color coordinate the clothes he wore. Each night, he wanted them laid out, ready for the next day when he prepared to leave for work, something Patty became quite adroit at handling. She would ensure the colors of the pair of socks matched, and help coordinate them with his various pieces of attire.

Laying flowers on a veteran's grave site.
Courtesy of The Milford Daily News

In late September, Fr. Connors called to inquire how the Herons fared and before the conversation ended, accepted an invitation to join them for lunch on Wednesday. Azelia wanted to express her thanks to the priest for his personal hand in aiding her husband during a crucial period on the battlefield. A religious person, she felt a meeting with Fr. Connors might provide spiritual healing for both her and Heron.

Seated in their living room, Fr. Connors talked about how they met unexpectedly on the battlefield where he performed last rites. He referred to Heron's rehabilitation and the productive life he led as husband and father, lauding him for his great courage and endurance throughout. "Your faith remains strong and I commend you for your strength and persever-ance, which serves as a great lesson for others wounded in battle."

"I felt like I died that day you found me, Father, and that God, for whatever reason changed his mind about killing me, deciding instead to give me a second chance."

"What a wonderful way to look at it. I agree He has his reasons for how things turn out, though at times it remains a mystery."

"There's an aesthetic element to blindness. For example, I once judged people by their looks and now realize judging people that way is shallow and absolutely meaningless. These days, I go strictly by what's inside and how people make me feel."

"A day of blindness might do us all some good." Fr. Connors surmised.

"It sensitizes your ears to such things as music and makes you a better listener because it frees you of visual distractions."

Azelia said, "On our drives to the beach, he knows how close we are sooner than anyone because he can smell the ocean from miles away." After a pause, she added with a chuckle, "And he never has to drive us there."

Lowering his voice to a confidential tone, the priest looked across at her with undeniable admiration. "People don't just talk about one hero in this family. Your name comes up frequently in that regard. I don't know if Larry told you but he made it clear that day on the battlefield he though of nothing but you, and might not have made it without that."

"*Would* not have made it," Heron corrected.

Connors continued. "Everyone gave up on him but he kept fighting to stay alive because he just had to get home to you."

When Fr. Connors described how his men drew straws to choose who would put him out of his misery, tears fell and she reached for Larry's hand. "Yes, I heard they were convinced he inevitably would die and wanted to end his suffering, and I would have died if he did not come home."

"I always thought that your prayers helped me, Father," Heron said. "Your presence that day meant a great deal and it helped to have a friend close to God as my advocate."

"I administered the last rites to so many young men in the war." Connors paused, staring into space. "I can count on one hand those who survived. I only hope I did some good for the souls of the unfortunate."

His voice suddenly took on a more euphoric note. "That brings me to another subject. On Sunday, November 12, I plan to hold memorial services for members of the 9th Division. The 9th lost 4,581 brave men who, like you, did not think twice about what the country asked of them. They lived their lives dedicated to their own faith, our faith, praying together with a common purpose. The world owes you more than it can ever repay. You risked your own live to save others and there's no greater tribute anyone can pay mankind." He laughed at the lack of a better word. "That includes womenkind as well, of course."

"Memorial services. Such a great idea, Father. No one should ever forget those men," Azelia said.

"I plan to hold services every year at this time and expect soldiers of all faiths to attend. I hope to see you both there, this and every year."

For a moment it seemed as though his mind left the room but it quickly returned. "It's funny. On the battlefield a man's faith never mattered. Soldiers of all faiths would congregate around any chaplain to pray. Like the Jewish chaplain who sent for me and when I arrived found him lying in a pool of his own blood on a battlefield in southern France. Knowing full well he was dying, he asked for my blessings, not absolution mind you, just my blessings." Connors' eyes glistened. "He died a strong, dedicated Jew wanting my blessings and nothing more. We should all have such faith, we must, and we must always stand up for it."

The room grew silent except for the ticking of a grandfather clock in one corner. "I'd like for us to pray together," Fr. Connors rose from his chair. "Stay right there." He knelt on the rug and reached out to join hands. For a long beat they sat in silence in a tight circle while a current born of emotions passed through their hands like electricity. In the background the grandfather clock clicked onward as Fr. Connors broke the near silence to lead them in simple prayer.

"We must keep our faith in God strong." He then spoke of love, love of family, country, and one another. He prayed for the souls of all who sacrificed their lives for their brothers and sisters then ended by thanking God for His divine blessings. While he spoke, Heron felt power and energy radiating from the priest through his hands to Azelia and back again, as though a spiritual presence occupied them and filled the room.

After Azelia served lunch, they sat sipping coffee and the priest lightened the mood by telling humorous battlefield tales. While the men continued to swap stories, Azelia left to refill the coffee pot and to slice a chocolate cake she prepared earlier that day. As he shifted to a more comfortable position, Fr. Connors took note of the pillow beside him with an intricate pattern woven into it. "What lovely colors," he said, holding the pillow with a floral design out in front of him. "Didn't know you did crewelwork," he said to Azelia.

"I don't. That's Larry's work."

"You're kidding. Larry, this is beautiful."

"Well thanks, Father but I'd appreciate it if you didn't spread it around."

They all laughed.

"No need for shame. Would you believe I learned to knit in the Army? Made myself some beautiful scarves and sweaters."

"I can't believe what I'm hearing," Azelia chided. "But somehow it makes me happy to hear it."

The conversation reverted back to Connors time as a chaplain and Azelia said, "We read a story in the Gazette of how you won the Silver Star."

"Oh that. It was nothing."

"Azelia read to me how you rescued some soldiers. No one dared pass through a minefield to get to them, so you went right in and pulled them out. I'd say that was something."

"They called me brave, but I knew I was in God's hands the whole time." A pensive smile crossed his face. "You know, when I learned I would receive the Silver Star, I never gave it a second thought. In fact, they held a parade to present it to me but guess what?" Before anyone could answer, he continued, "I was so involved with administering to the spiritual needs of the 9th that I totally forgot to show up. Can you believe it? I completely missed my own parade." He chuckled aloud. "How embarrassing is that?"

"Guess you were answering to a higher calling."

"But more thrilling was the day I captured twenty-five Germans."

"Now, that's a story I can't wait to hear," Heron said.

"A rather bizarre one at that, but I love telling it." He lowered his coffee mug to the table and sat back. "A soldier from the 9th Infantry was driving me through Aachen, Germany over a treacherous road full of holes and craters. We passed fields dotted with charred bodies covered in dust and rubble. Skeletal sections of buildings stood among the ashes, mounds of brick and blackened debris, ready to collapse. Fires still burned along the sides of the road and black smoke billowed to vanish into the dark of night.

We drove until we came to a clear stretch of road with no bodies or debris, just a barren strip laced with fog so thick we could only see three feet ahead in the distance. I can still hear the sounds of the driver shifting down gears and the engine taking on a loud yet comforting rumble as we moved cautiously forward."

"After a while I began to doze and apparently the driver became somewhat mesmerized as well because when we rounded the next corner, he nearly plowed into a column of German soldiers just as startled as we. They dodged to one side of the road to avoid getting run down and my driver instinctively pulled the jeep to a halt to avoid hitting anyone. Well the surprised Germans just froze in place and not one of them raised a weapon or said a word."

"I asked the driver, 'What'll we do?'"

"He answered, nervously, 'Yell something at them, sir, as loud and brassy as you can!'"

"In German I commanded, 'Put down your weapons and surrender.' That was the first thing that came to mind. 'The war is over for you men.'"

"To my amazement, the officer leading them saluted me then approached the jeep and handed over his weapon. Then one-by-one they each came forward to drop their weapons into the back of the jeep. When the last man laid his weapon down and stepped away, all raised their hands in surrender. I couldn't believe they were surrendering to me, a chaplain."

"You're kidding," Heron laughed.

"Thank God they weren't SS or they would have killed us on the spot."

"What does SS stand for?" Azelia asks. "I've always wondered."

"Schutzstaffel, a wild, vicious cadre of animals whose role was to starve, brutalize, torment, torture and murder helpless civilians. They wore the same uniform as storm troopers: gray jackets with swastika armbands except their caps were black and fixed with a silver death's head badge, and they wore black ties. "

"When I led the prisoners into camp, the men couldn't believe their eyes. Word spread quickly that I captured twenty-five Germans, and that led to quite a ribbing. 'At least I take them alive,' I told them."

CONNORS DROVE home late that afternoon pondering what might have been. Heron had been born gifted beyond a young man's dreams, gifts consumed that fateful day on the outskirts of Cherbourg. *Yet I never heard him complain.*

He overcomes major obstacles every day of his life and continues to endure painful operations. He marches in parades with battered head held high. With the loss of his athletic skills, he discovered a hidden talent – a beautiful voice for the enjoyment of others at hospitals, churches, and

*civic events. He speaks encouragement to young people and maintains a
love of sports, though denied the opportunity to watch them. Instead of
sitting in a home wasting away he stays active, socializes with old friends
and makes new ones.*

Aloud he said, "And Azelia is a woman blessed with great beauty,
courage, strength, and devotion who neither complains nor expresses
bitterness. She's been his crutch, his beacon. And both maintain a strong
belief in God and a sense of values. What a great country! What fine
people!"

Fr. Connors suddenly heard the sound of his own sobbing.

ABOUT TO DOZE off, Azelia presses the button to adjust her bed raising her to a sitting position. The view out her window seems brighter.
Soon morning will arrive and the feeling that she may not make it to sun-
rise looms stronger than ever. She must continue where she left off and not
think about dying.

Though their income during this period of their lives did not allow
for extravagant living, it adequately addressed their needs, thanks in part
to the support system afforded by a prosperous town, and partly to the
generous legacy passed on by the thoughtful Drapers. Almost everything
needed could be found within walking distance of their house including
a pharmacy, hospital, police and fire departments, a pond, parks, library,
grocery store and bus transportation.

The townspeople remained grateful for Heron's sacrifices and continued to treat him and his family with respect and admiration. Attitudes
would change with time but during the first decade following the war,
people did not forget the sacrifices he and his family made for the welfare
of the people living there.

The following article appeared in the Milford Daily News on Tuesday, November 19, 1946: *At a well-attended meeting lasting two hours last
night, the Milford VFW unanimously voted to present Lawrence J. Heron
of Hopedale, a gift of $500 for Christmas. The drive to raise the money
began last July with letters mailed to veterans in Milford and Hopedale.*
The article ended with a notation that the American Legion kicked off a
drive for the purpose of doubling that amount to a thousand dollars, about
half the average annual salary that year.

On December 21, 1946, the Army awarded Heron his discharge.
Ironically, the discharge papers listed his eyes as blue and he left with a

severance check in the amount of $154.77 and the Purple Heart. He quick-
ly settled into life as a civilian, making his own way to and from his job
at the factory each day. In winter months when snowstorms hit, he would
head out with a shovel, and worked around the house in the summer, even
performing maintenance on the family car. On weekends, Heron fished at
the local pond or spent the day on the golf course with friends. Most vacation days Azelia drove the family to Marshfield where he'd swim in the
ocean or dig for clams.

NORMA RIPANTI married soon after the war, changing her name to
Norma Thurston. As she entered Azelia's kitchen to help clean up following a Halloween dinner party in late October, she asked, "Do you know
what your sweet little daughter just said to me?"

"Nothing bad, I hope."

"She pointed to her own eyes and said, 'Daddy uh-uh.'"

"Oh yes," Azelia laughed. "She says that sometimes when he takes
off his glasses. She's very aware that her daddy is different and that he
has no eyes. Normally, she's a bundle of energy, getting into everything. I
can't keep up with her, but when she's around him she calms right down.
She will take his hand the minute he stands up and lead him around like a
little seeing-eye dog."

Norma shook her head and smiled. "This family amazes me, especially you. You're so strong. I have always admired your courage."

"Oh, but how about you? Graduated from Temple in '42. A master's in physical therapy, and then off to join the Army."

"Yes, but I ended up safe at Walter Reed for two-and-a-half years
until my discharge this year."

"You don't think that's something? You rehabilitated injured soldiers. You became an officer for God's sake. You accomplished something
with your life."

Norma raised a brow. "Do I detect some...some discontentment
here?"

Azelia blushed. "I'm content with my life. Oh, sure, now and
then I wonder what goes on outside Hopedale, but I have a life, relatives,
friends..." She trailed off, moving away in her mind then swiftly returned.
"When I think about it, the only thing I haven't satisfied is my curiosity. It would be nice to travel. I'd love to see Italy, but I know once I got
there I'd begin to miss everything and want to hurry back. It doesn't get

any better than this, just new and different perhaps. But I can live without change."

Norma turned her forthright eyes on Azelia and said, "How did you become so damn wise, Mrs. Heron? It takes most people years of travel and worldly experience to come to that very same conclusion."

Heron fishing with a friend. *Courtesy of Larry Heron, Jr.*

ON A LOVELY spring morning in 1947, Heron and Azelia went for a stroll arm-in-arm through town. The day remained sunny and bright, the air clean, the sky a vivid blue with temperatures in the low 70s. Ordinarily, it would take but five minutes to pass through Hopedale, but nearly everyone they met eagerly engaged them in conversation, slowing their progress to a crawl. As their walk continued, Azelia noted Heron displaying less than usual patience and rarely spoke to her.

Today's walk differed from any other because Heron planned a surprise for Azelia. Following his instructions, she turned them right at the corner of Hope Street and followed it for a block. Then he asked her to stop in front of Adin Ballou Park. In the right-rear corner of the half-acre park, stood an imposing ten foot bronzed statue of Adin Ballou before the only remaining remnants of the settler's original farmhouse, which included a doorstep and a boot scraper.

"So tell me," Heron asked. "What do you think of this part of town?"

"It's one of my favorites. The homes are lovely and it's a half block from the library where I spend much of my time. Why do you ask?"

"Tell me what you think of this next house, the one abutting the park."

"The one to our right? It has a lovely yard. No one can ever build on the park side. It's almost like an extension of their yard."

"Make that our yard," he corrected.

"Our yard? Larry, what are you saying?"

"Yesterday, my boss told me the house became available and that we are first in line if we want to live their. I told him we'd take it."

"Oh, Larry!" Azelia squealed with delight.

"Our old place worked for us fine before I joined the army, before our marriage and before our family grew. But since then, I could swear the rooms have shrunken and the ceilings dropped, even the windows seem smaller."

"I thought you couldn't see."

"It just feels tight living there. We need more room, don't you think?"

A Draper representative waited at the front door to take them on a guided tour of the first floor to show Azelia a huge kitchen and dining area, living room, family room, two bedrooms, and one-and-a-half baths. On the second floor he pointed out two more bedrooms and a full bath. They completed their move to Hopedale Street in the summer of 1947, much to Azelia's delight.

With the war at an end, the country redirected resources and technology toward developing products for a peaceful society and soon came television, frozen foods, stereos, 45-RPM records, and chickens tidily wrapped in plastic. Doctors stopped making house calls and mail delivery in Hopedale dropped to once a day.

And on January 1, 1949, Carol Heron entered the world.

IN THE SUMMER of 1951, the Disabled American Veterans headquartered in Chicago, distributed an account of Heron's story to newspapers across the country: *the only living veteran in the history of the DAV to have a chapter named for him...overcame insurmountable handicaps to lead a normal life...outperformed sighted peers as a quality control inspector at the Draper Corporation...survived injuries that would have*

killed an ordinary man...endured many complicated surgeries...serves as
an inspiration to fellow members of the DAV and to the handicapped.

Heron interviewed during a radio broadcast of his life.
Courtesy of Larry Heron, Jr.

Heron's notability grew until celebrities and high-ranking offi-
cials sought him out for photo-ops or to appear with them on television.
In 1951, a summary of his life was broadcast nationwide from Worcester, Massachusetts during a half-hour radio program. Then on September
9, 1952, the Heron family gained a new addition, this time Lawrence J.
Heron, Jr. entered the household where two men would soon answer to the
name of Larry.

CHAPTER TWENTY-ONE

FAMILY

THE FOLLOWING summer, the Herons rented a place in Marshfield, Massachusetts and took note of a cute cottage for sale a few blocks away with a lovely ocean view. Although in need of repair, it would not take more than a few fresh coats of paint, new appliances and furniture to add sparkle inside, and a bit of landscaping to spruce up the outside appearance.

The price, double what they felt they could afford, loomed as a deal breaker when something happened that on the surface appeared like an answer to their prayers. When Heron described the property to a close friend who lived in Hopedale, the friend replied, "I always wanted a place on the water. My wife and I should take a look. If we like it, perhaps we can go in with you."

Azelia fell out of favor with such an arrangement. "It's a good way to lose a friend," she told Heron when alone. But Heron really loved the ocean and whenever he could get to it, swam like a marlin and so she eventually relented, deciding he should not be denied. On the next weekend they journeyed back to Marshfield with the friends who took one look, and without hesitation, declared their intention to become joint owners. "We love the place but could never afford it alone, so if you want to share the cost of ownership, I think we can put a deal together." After reviewing the details, his friend said, "Okay, let's do it."

So the following week the Herons put down a deposit to hold the property then returned home to contact their friends. "We can all sign the necessary forms and get them notarized." Heron told the couple, "The realtor plans to call us tonight with a closing date, meanwhile we'll add up the costs and split them equally."

"Gee," his friend responded, "I wish you'd spoken to me before putting making a deposit. We went over our finances last night and decided we can't afford it right now. I meant to call and tell you but something came up."

"But I thought we agreed."

"I'm sorry Larry. We changed our minds."

When Heron filled Azelia in on the conversation, disappointment weighed heavily on him. "Let's go through the numbers again," she told him. "It looks like a stretch right now but we both expect pay increases this year, so let's see if we can make it work. I would rather we owned it by ourselves than put up with the hassles of joint ownership."

It turned out the best decision. Clearly, owning property with another party would not end happily, they realized. Upon deciding to buy it themselves, the Herons focused on fixing up their second home, delighted at the opportunity to make changes as they saw fit, with no interference from co-owners.

Azelia loved helping her father in the garden and learned from him the secrets of why Italians possess green thumbs. Flowers and vegetables alike flourished under her skillful and loving care. She'd done so well with their new house on Hopedale Street that visitors could no longer distinguish where the park property ended and theirs began.

The Herons traveled to Marshfield almost every weekend transforming the rough-shorn yard surrounding the cottage into the envy of their neighbors. Her brother, Guido applied his artistic talents to painting the outside of the cottage then spent his vacation constructing a gorgeous stone fireplace in their spacious living room. When finished, it added value to their dream cottage and the warmth of a fireplace helped extend the season well into the fall of each year.

BY NOW, most Hopedale area residents regarded Heron a living symbol of everything the American soldier stood for and he became an icon in the minds of wounded veterans. Admired for his heroic actions and courageous rehabilitation, he served as a role model for young people, especially the handicapped.

Locals wanted to help in any way they could, like driving him regularly to the Bright Oaks Club on Mendon hill to treat him to free drinks. At Bright Oaks, a social club where a blind man could socialize with sighted people on equal footing, his beer glass never emptied, and he seldom reached for his wallet.

Casual conversations and friendly arguments usually won the day at the club and Heron loved the gregariousness, the smell of beer, the laughter, the loud voices and soft music always playing in the background. Here he could listen to discourses on everything from politics to how to best fertilize the garden, all while drinking beer and playing card games.

The club kept sets of Braille-marked playing cards on hand just for Heron and in time people found him unbeatable at poker. When he played gin, he added up his points faster than his opponent. With no distractions, he could focus his mind on the game, count cards and remember every one previously played. Competing with him was like playing against a computer.

And all the while, Heron's taste for beer grew stronger. It helped him forget the handicap thrust on him by the hand of God. More often than not, by the time someone dropped him off at home, he struggled to reach the front porch then needed help to make it up to the second floor so he could crawl into bed. On one such occasion, after Azelia drew the covers over him, he thought he heard her soft sob.

The next evening, a Saturday night, he walked past Azelia and into the kitchen where he opened the refrigerator to remove a bottle of Budweiser. "Care to join me, he asked."

"No thanks. I'm still working on a glass of lemonade."

He popped open the beer and set it down on the kitchen table then took a seat while waiting for Azelia to sit down opposite him. "Do you remember that song I wrote for you after our...our one and only breakup?"

"Yes." She would never forget. They had been sweethearts all through high school, and as part of the courting ritual he would drive her around town singing songs he'd created just for her, songs she liked so much that she didn't believe him when he first told her he made them up. She would never forget the one he sang to her not long after the breakup. When his father died, Heron invited Azelia to the funeral but she failed to show, which hurt his feelings. Angry that she did not support him when he felt he needed her most, he decided not to speak to her.

It happened early on in their dating cycle, before she got to know his family very well. Livia, Azelia's mother, thought it improper for a young woman to attend such an event when she barely knew his family. Where Livia grew up in Italy, people women would not date without a chaperon present.

Heron did not speak to her for over a month and might have continued longer if Azelia did not attract other young men who began queuing up to ask her out once word spread of their lack of commitment. It came to a head when a close friend approached Heron whose friendship he valued, announcing he planned to ask her out. "I was wondering if you would mind?" Heron replied, "Hell yes, I'd mind! Azelia's the woman I plan to marry."

Fearful he might risk losing her, Heron called that same night to invite her out for a drive. When she accepted, he parked in a remote spot and told her he wrote a song just for her that he proceeded to sing. She loved the song and listening to it went down as one of the most romantic moments in her life.

So, as they sat quietly in the kitchen separated by the ominous bottle of Bud, he looked up at her and said, I never wrote the words down anywhere and wish I could recall them now. "You didn't by any chance jot them on a piece of paper somewhere?" Then he answered his own question. "No, of course not. Why would you after all these years?"

"No. I never did," she answered.

"I'd love to sing it to you one more time."

She turned from him so that he could not see her smile then turned back feeling silly, realizing he couldn't see. "But I remember the words, every line."

"You do?"

"Sure." She didn't hesitate a moment. *"Sweetheart, I'm sorry that I made you cry, sorry for each little tear in your eye. I was mistaken, my whole heart is aching, please tell me you won't say goodbye. Now our dreams were broken, but soon they will mend. Kiss me again. Hug me again. Sweetheart, I'm sorry."*

The silence weighed heavy for what seemed a full minute before Heron spoke again. "Sweetheart, take a good look at this bottle of beer because it's the last one you will ever see me drink." It was his way of telling her that he hated himself for making her cry on that night he returned home drunk, and that it never would happen again. Heron always kept his word. She never saw nor heard of him drinking alcohol from that moment on.

Around that same time, the Draper Corporation experienced its first layoff. Though the demand for looms had fallen sharply and the company's profits continued to erode at a fast rate, the company continued living up to its image of caring. Rather than downsize, Draper announced it would keep everyone working, but only three days a week.

ON OCTOBER 30, the Herons received an invitation to attend the annual memorial Mass for the men of the 9th Division lost in World War II to take place at the new Immaculate Conception in Worcester, Massachusetts. Its pastor, Rev. Edward T. Connors, directed the construction of this brick and stone church located about a mile from its predecessor. The note attached to the piece came from Fr. Connors requesting Heron come prepared to sing during the reception.

The invitation arrived with an enclosed article written by Fr. Henry Murphy under the synonym "Canon Pepergrass." The elderly Fr. Henry, a close friend of Fr. Connors, described *Connors Coffee Shop* as a place on the battlefield with a pot always brewing for any GI to enjoy at any time, and where friendships flourished while great bull sessions came to pass. Peppergrass dispersed bits of humor and nostalgia throughout the article intended to set the stage for an interesting and exciting event intended annually for members of the 9th Division, a project Fr. Connors would faithfully perpetuate every year for the rest of his life.

Over 450 people of all faiths attended memorial Mass that year, with parents of many of the deceased traveling from as far away as Pittsburgh. On Saturday night a special get-together allowed husbands to introduce wives and children to old army buddies. An hour later, Heron took front and center to sing, and by the time the social gathering ended, he and Azelia had conversed with nearly everyone in attendance.

The emotionally charged services represented a solemn dedication to the honor of veterans and their families, the nation as a whole, and the high moral purpose and idealism that had motivated the nation's call to arms. That evening after an elaborate dinner, Heron sang a medley ranging from *Battle Hymn of the Republic to Prayer of Thanksgiving*. As always, the Herons later met new people who quickly became friends. When services ended, the couple drove home promising to attend these events in coming years as often as possible. Each year thereafter, attendance at the memorials continued to increase, doubling within five years.

AZELIA SCANNED the Boston Globe each morning, and on December 24, 1954 came across an article she could not wait to share with Heron, "It says that Dr. Joseph E. Murray performed the world's first successful kidney transplant at the Peter Bent Brigham Hospital, and that it's the first organ transplant ever performed."

"*Our* Dr. Murray?"

"The very same."

"Honey, that's wonderful. I always felt he'd achieve greatness in our lifetime."

"He said he was influenced by Dr. Brown's cross-skin graft of a pair of identical twins in 1937, and that his organ transplant was between identical twins as well."

"Just like him to spread the credit around."

ON MAY 5, 1956, Debra Heron joined the family and like any proud new father, Heron doted on his new baby girl. His children's presence and their laughter filled him with great happiness and joy. Through his children, he could vicariously reclaim pieces of his lost life.

As each child reached school age and began to encounter everyday challenges, she or he would approach Dad for advice, and he would try his best to help them seek out and find their own answers when possible. He wanted them to learn that they possessed the strength to stand on their own. Heron attended all school functions, taking pride in their achievements, and gave his complete attention whenever one wanted to talk, or his arm to lean on when needed.

It upset all of his children that he could not see them or watch them perform in school activities. They would take the time to describe their clothes, schoolwork, new devices, everything in great details to help him visualize and draw comparisons to similar objects he had seen in the past.

People proclaimed young Larry a chip off the old block and told him that Patty's eyes matched Azelia's or that Debra had her smile. It delighted Heron when people told him that Larry took his looks and that the girls inherited theirs from Azelia. "I would have been very unhappy if the opposite were true," he would tell them. No matter how often he heard a description of one of his children, he found no way to form an image of the whole person and therefore would never truly know the looks of any one of his children.

A LETTER FROM Dr. Vance Bradford arrived in April 1961 thanking Heron for allowing him the use of photographs taken just prior to the plastic surgery he performed in England. Dr. Bradford enclosed a photograph of his visit to Stonehenge upon a return visit to the hospital where he performed plastic surgery on Heron.

The doctor also enclosed copies of the paper he published in the May 1961 Oklahoma State Medical Journal entitled "Burns in Atomic Disaster." It described medical problems likely to result from an atomic bomb blast and included four photos of Heron to illustrate how victims of an atomic blast might appear.

The gruesomeness and poor quality of two of the four photos prevents showing them, but one can readily imagine the severity of his burns when the doctor used them to depict the extreme horrors inflicted on survivors of an atom bomb blast.

In his letter, the doctor wrote: *"We had several who lost one eye but you were our only patient who got both knocked out. I remember you very well, dictating letters to your wife, walking about the hospital grounds, etcetera. One of our patients who wrote to us asked about you. He had extensive electrical burns. I took the liberty of giving him your address."*

Dr. Vance Bradford visiting Stonehenge. *Courtesy of Larry Heron, Jr.*

BY MID-SEPTEMBER, eighty-year old Emma demanded admittance to a nursing home over Azelia's objections because she now required constant care. "Your hands are full with the children," she insisted. "If I stay, everyone will suffer. No. It's best this way." Heron's mother passed away in her sleep at the nursing home three months later on December 23, 1961.

"Why just before Christmas?" Heron wanted to know.

"Many people become depressed at this time of year. She loved the holidays so much, but lost the joy of living. Though mentally alert, her body quit on her, taking with it the quality of life and her desire to live any longer."

"Yes. I guess that's right. She stopped enjoying her time with the kids, the singing and the laughter eluded her." The pain went deep because he loved her dearly. "Perhaps we shouldn't celebrate Christmas this year," he said.

"You know she would have wanted it. We need it more now than ever."

"Yes. I suppose you're right, though it could turn into an Irish wake."

Heron at #158 General Hospital, Salisbury, England in 1944.
Courtesy of Larry Heron, Jr.

MORE PEOPLE than usual crowded into the Heron residence on Christmas Day that year. The Johnny Milans and the Jerry Dees arrived early. Olga and Fred Bresciani, Amelia and Jimmy DiSabito, Vera and Joe Pantini, and all brought their children. By the time Ed Kalpagian and Pete Ferrelli arrived, Larry anxiously stood by to greet them. "You guys sure are a sight for sore eyes,"he told them, which lightened a darkened atmosphere and brought laughter.

Ed relieved Olga at the piano, playing Christmas carols while guests sang along. The eggnog flowed and folks properly toasted Emma. *Unbounded love and kindness...a great lady...hell of a cook...always there for you, especially for Larry...they were very close.*

In keeping with what long ago became a Christmas tradition, Johnny Milan delivered his rendition of *Rudolph the Red-nosed Reindeer.*

When he finished, everyone drank a solemn toast to Rudolph and his reindeer friends. Jerry Dee and his wife departed early, bidding all a "Merry Christmas and in the New Year, may your right hand always be stretched out in friendship and never in want!"

Just as holiday celebrations took place at the Heron residence each year, the summers included vacations at Marshfield, fishing, clamming, golfing, backyard barbecues with neighbors, and swimming in the Atlantic Ocean. Despite blindness, Heron would swim like Johnny Weissmuller, farther out from the beach and for longer periods than his guests. The only help he requested were directions toward shore when he felt ready to call it quits.

CHAPTER TWENTY-TWO

CHANGES IN ATTITUDE

THE YEARS sailed past with alacrity and before Heron knew it, Debbie received an invitation to her high school prom. She seemed excited enough when describing her dream date to him but Heron found her melancholy whenever he talked to her about the clothes she planned to wear. It saddened her that her father would never see her in a gown. In fact, he would never see her progressing to become a woman, no longer just his little girl. Never throughout her life did he look at her face or see that she had developed a figure. The fact that he's never laid eyes on any of his four children disturbed young Debbie perhaps as much as it did him. *It's so unfair.*

She could list an entire range of absurdities accompanying his blindness. At Christmas he never got to fully appreciate the tree or the colorfully wrapped gifts, nor could he see the joyous looks crossing the faces of his family when it came time to opening their presents. He could never view a colorful sunset, nor look upon a snow-covered landscape.

But tonight, damn it, he will find out what I look like! Well, almost.

When the time came to make an appearance, she looked in the mirror one last time trying desperately not to cry and spoil her makeup. If only he could see her light blue taffeta gown, the polish on her nails, the high heels, and hair swept back and gathered loosely in the latest fashion.

"You look beautiful," Azelia told her as she came down the stairs from her room. "Yes," her father confirmed. "You'll knock 'em dead tonight."

She walked straight up to him and took his hands, pulling gently to get him to rise from his chair. "Stand up, Daddy," she told him. With that, she took a deep breath and placed his hands on her bare shoulders then slid them to her sides. "You can't see my dress," she said. "But you can feel it. It has a slight blue tint, and it fits perfectly," she said. "Go ahead." She guided his hands as she described every last detail from the spaghetti straps to the taffeta around her waist, the scoop in the back, and the flared skirt.

"Fits perhaps a bit too well," Heron quipped. No matter how grown, she was still his baby girl. "That guy better have you home no later than one o'clock."

"Dad! This is a prom."

"My little girl has grown up. You look lovely," he said, giving her a peck on the cheek.

"Thank you."

"No." He hesitated. "Thank you. And you smell of jasmine." He could hear Azelia snickering in the background, or could it have been a sob?

"Remember, one a.m. at the latest!"

"I'll try." Debbie felt happy because she sensed her father was both surprised and pleased. That little episode helped him visualize his daughter grown up and neatly fitted with a prom gown. It perhaps helped him gain a better understanding and feel of what his children looked like.

After she left with her date, Azelia said, "Trust me, sweetheart, your daughter is blonde and beautiful. All your children are lovely. On a scale of one to ten, we produced elevens."

"No false modesty here," he replied, great joy registering in his voice. "I've never felt prouder!"

THE MOOD of the nation shifted and as Heron would learn, not all for the best. Arriving home from work on a Monday in early 1967, he announced, "North American Rockwell bought the Corporation out from under Ben Draper."

Azelia shook her head. "I can't imagine the place not owned by a Draper."

"The market's gone dry. What business exists, spare parts and such long since moved south to Spartanburg, South Carolina."

"I'm afraid time has passed by the textile industry. The first Draper looms produced over a hundred years ago look no different from today's, and the old ones never breakdown."

HERON'S CHILDREN thought of their father as very special, and grew proud of his many accomplishments. As a fairly captive audience, they could depend on their dad more than most kids they knew. He spent more time with them, digging for clams at the shore, fishing, golfing, swimming, or just hanging out and talking with them like one of the guys.

They marveled not only watching him patiently fashion a fishing fly but actually fly-fishing, catching more trout than they could While other fathers seemed too busy working late, traveling or pursuing promotions, their handicapped father stayed in town and focused his attention on them but not in an overbearing way.

Memorial Day, Veterans Day, Independence, Flag or Armed Forces Day, no holiday passed without a newspaper article or two about Heron's achievements, rehashes of his athleticism, service and sacrifices, his courage in the face of battle, but mostly on the subject of strength, perseverance and survival.

Heron the war hero gained as much celebrity as the Heron who once dominated sports pages. So he became more conscious of his looks and began paying more attention to his dress. To that end, he relied heavily on family members to ensure he wore the color coordination of his clothes: white or blue shirts subtly matched with his gray, black, or dark blue designer suit; plain gray, black, or dark blue socks to complement any suit. But he needed help with minor details as well, like when it came to pairing items. "What color are these socks?" he asked Carol the night before the 1968 Memorial Day parade.

"Brown," she said then quickly added, "with blue flecks – to match the red stripes in this shirt."

"What?" he exclaimed.

"And the red diamonds on your socks that go with this pink tie."

"Oh pipe down," he said.

"Hey, watch it or I might match one red sock with a white."

He liked when his children played pranks because it showed that in no way did he intimidate them or arouse their pity. They never knew him any other way so blindness never shocked them or caused the least bit of embarrassment, and that suited him just fine.

LATER THAT year, at a reunion held in New York for former members of the 87th Chemical Weapons Battalion, John Sears passed out copies of several articles he wrote for newspapers serving towns south of Boston.

One article explained that the 87th landed on D-Day armed with mustard and phosgene gas. In it, he told how the men wore special-issue fatigues dipped in wax with brown patches on the shoulders that would turn pink when exposed to poisonous gases.

As the war progressed, shells filled with poison gas did turn up in German ammunition dumps along with thick rubber suits made for gas handlers. Field Marshal Erwin Rommel indicated the Germans did not use poisonous gases for only one reason, they failed in attempts to develop a mask that could protect horses. A severe shortage of oil forced them to rely on horse-drawn wagons to transport their ammunition, food, and medicine, and the Germans could ill afford to lose their animals as a consequence.

The Americans took Cherbourg at the end of June and in one warehouse came upon a cache of Wehrmacht bootleg booze. Needing room on their trucks for gas and booze led an unnamed officer to approach Sears with an order."All right, Sears, I want you to get rid of that gas."

"Get rid…? Where am I supposed to put it?"

In the finest of military traditions, the captain simply glared at him and said, "I just gave you an order!" code language for, "Just do it! I don't want the details or take responsibility in the case of repercussions."

Sgt. Sears pondered how to dispose of the poison gas. Ammo dumps would not touch them fearing gas shells far too dangerous to keep on hand. Faced with no other choice, he took the only avenue left open to him.

Over the next few days his crew collected all the mustard and phosgene-filled shells from each of their four companies then searched until they located an open field with cattle grazing on it, which proved it unmined. With no one within miles to witness, he ordered his crew to dig a deep trench twenty-feet long and three feet deep where they buried the shells. As far as anyone knows, the poisonous gas shells remain buried there to this day.

Sears wrote many articles about the experiences of the 87th and sent them to all the members he could locate. Azelia would read them to Heron, including one disturbing article dealing with Hitler's death camp, Nordhausen, a story of deprivation and horror that left her on the verge of tears.

As Sears came within a mile of Nordhausen, the gagging, putrid stench of human death hit his stomach unlike any odors he would ever again encounter on or off the battlefield, a smell that would stay with him throughout his lifetime.

When he reached sub-camp Dora-Mittelbau, he came upon a hundred people still alive, their emaciate bodies and hollow eyes telling a tale of man's inhumanity to man. Dressed in tattered clothes and barefoot, none could have weighed more than 70 pounds, and in an open field outside the barracks he found bodies stacked six high on pallets set a few feet apart. About five thousand dead lay like skeletons with skin stretched over them in varying shades of white, purple, yellow, and green.

Unidentified soldier viewing bodies of victims.
Courtesy of John Sears.

Only a few dozen of the thousands of bodies belonged to Jews, the rest, records confirmed, included Poles, Czechs, Russian, French, Belgian, and Dutch. Inside the barracks, the unbearable stench surrounded inmates laying in beds like cardboard cutouts right where they had died. The few found barely alive suffered from typhus, were crawling with lice and covered with sores.

The Germans had forced these prisoners to manufacture the V-2 missiles they then launched against cities like London. Rocket assembly took place in multiple caves carved into the mountain that were interconnected by long tunnels used for the transportation of the missiles by railcars. Enslaved inmates built and transported these new weapons of mass destruction for as long as they could, until overwork and starvation finally led to their deaths.

The Americans who freed them unintentionally killed a few more by feeding them food instead of starting them out on teaspoons of water.

All ages and genders. *Courtesy of John Sears*

The Americans rounded up German civilians from nearby towns and forced them to bury the dead. Their denials of knowing what had taken place at the camp fell on deaf ears, for the smell alone loudly proclaimed to anyone within a hundred miles exactly what went on here for a number of years.

When Azelia finished reading the article, to Heron, any doubts he harbored as to whether he had suffered in vain were greatly diminished. Freedom, he decided does not come without a steep price.

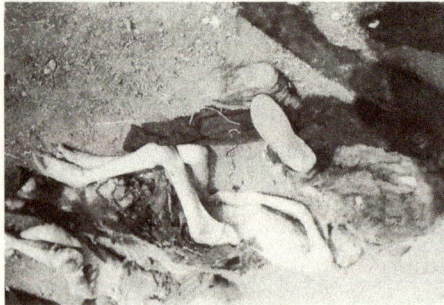

Never may we bear witness to such horrors again.
Courtesy of John Sears

IT STARTED with the Gulf of Tonkin, a body of water on the East Coast of North Vietnam, the staging area of the U.S. Seventh Fleet, the site that would lead to the escalation of U.S. involvement in America's next war. July 1965 found 80,000 U.S. troops stationed in South Vietnam that by 1969 would increase to 543,000 while 400 tons of bombs and ordnance would fall on Vietnam daily, the beginning days of the most vivid and memorable war in modern day history.

The longer the war continued to rage, the more the nation rebelled against US involvement. Faith in American leaders and American power fell as opposition to the military increased dramatically since World War II, and veterans suffered the scorn of advancing numbers of anti-war protestors.

When the war would drew to a close in April 1975, the US had committed 2.6 million troops, unleashed many times the tonnage of bombs dropped by both sides throughout the entire span of World War II, and suffered 365,000 dead and wounded. Both North and South Vietnam, suffered an estimated five million civilian and military casualties by the time the war concluded, and its end marked the beginning of a series of events that sent Heron's spiraling into a deep depression.

CHAPTER TWENTY-THREE

DEPRESSION

ON A FRIDAY afternoon in November, as the sun moved high over-head and a cool breeze swept down from the north, Heron stepped off the sidewalk onto the grass to avoid colliding as one kid after another sailed past on their skateboards like dive-bombers peeling from the sky.

"Hey. Watch me," a voice called down from the top of the hill where the library stood. Heron heard skateboard wheels rumbling toward him on the rough pavement. The sound of a loose wobbly wheel identified the skater as Glenn. Billy's well-oiled board rolled smoothly past by comparison, always with a low hum so he could tell the difference. Walter's board gave off more of a rumble, like the sound of rolling thunder. Yes, definitely Glenn.

"Hi Glenn," Heron said, as the boy came abreast of him.

"Yikes!" Heron heard the board skid out of control and heavy footsteps as Glenn's feet hit the pavement and he pumped hard to keep from falling flat on his face. A final *whump thump* told Heron that the boy crash-landed on the soft lawn.

"You okay, Glenn?"

"Geeze, Mr. Heron, how'd you know it was me?"

"Just a lucky guess," Heron said, trying to suppress a laugh.

It rained Monday and Tuesday, and by Wednesday morning the sky took on a cool deep blue as thousands lined the parade route running through the heart of downtown Milford. Veteran's groups gathered, as did the Police and Fire Departments led by their chiefs. The local selectmen, state representatives, and politicians turned out as usual. Gold Star mothers of the Italian-American Vets, the DAV Colors and Contingent, the American-Armenian Vets, the VFW, and even a small group of World War I veterans rounded out the assortment of proud Americans marching in the parade.

Each time a flag group passed, veterans lining the route removed their hats and held them over their hearts in solemn tribute. Others tossed a salute, thinking back to the war they served in or the time each had served domestically. Clearly, everyone in these parts took their patriotism seriously.

A man in an American Legion uniform shouted from his wheelchair parked on a sidewalk, "Here comes Larry Heron," The named we gave our DAV chapter."

From the sidewalk in front of Billy Focus' Diner, another voice shouted, "Hey Larry!" Heron recognized the voice of John Jakes. Without losing a step, the blinded veteran called back, "Hey Johnny. How you doing?" Jaws dropped in amazement, even locals already familiar with Heron's ability to recognize a person by the sound of voice stood amazed.

As Heron reached the corner of Main and Central, he heard a commotion and the officer guiding him told him to stop. A group of protestors tried to block the intersection and a fight broke out. Chants of, "murderers, baby killers, and stop the slaughter" reached Heron's ears.

"What's going on?" He asked.

"Some jerk burned the American flag in the middle of the street. Cops are trying to remove him." A half-hour later, the parade resumed, but hearts and minds of the marchers now felt heavy. The country had never been this divided.

HERON BECAME upset when a weekly column in the Milford Daily News carried a quote by a woman claiming Hopedale's "blind people" filed complaints about children on skateboards menacing him. Heron knew the woman disapproved of skateboarding and felt she used him to form an excuse to launch her own complaint.

Azelia had never seen him this angry. "Blind people? That would be me!" he exclaimed. "I'm the only blind person in town. They're nice kids, doing what kids do. Now they think I complained about them."

"Well why don't you do something about it?" Azelia suggested.

"Let's phone the newspaper and set the record straight." She made several calls until an editor came on the line.

"And make sure you quote me," she heard Heron tell him before hanging up the phone. "Those kids never bothered me. Never came close. They're swell kids, all of them."

The newspaper printed a formal retraction the following day quoting Heron. "I had nothing to do with the article. No one spoke to me before printing it. If they had, they would know I have no problems with children or their skateboards."

Grand Marshal on Veterans Day.
Courtesy of The Milford Daily News

YOUNG LARRY turned seventeen on May 14, 1973, and began driving his father wherever he asked, and he especially enjoyed taking him fishing. The calm pastoral setting beside the pond allowed for some quiet time together. Golfing provided another enjoyable pastime at which Heron would amaze onlookers with long straight drives and putting skills better than a sighted person. On their way home from a day of golfing in the adjacent town of Mendon, Larry pulled into the parking lot of a grocery store to pick up a few items to bring home for dinner that night.

As they exited the car, a man angrily approached waving his arms and shouting, "What the hell do you think you're doing? That spot is clearly marked for the handicapped. Park that damn thing somewhere else." This outburst despite license plates clearly marked with a "V" to signify

Heron's ultimate right to handicapped privileges. Like any father protecting his son, Heron stepped from the car and curtly replied, "And what the hell's wrong with you? Are you blind?" He pointed to his sunglasses and added, "Can't you see that I'm blind?"

The man grunted words under his breath and stomped off. Heron read the skateboard incident and anti-war movements as signs of a major shift in attitudes toward veterans, one that did not strike him as a change for the better. Two days later, Azelia replayed a conversation she heard between two women at the post office who did not know she had doubled back for stamps. Earlier they overheard her ask the postal clerk if this month's disability check had arrived.

"I can't believe he's still getting disability payments," one of the women said.

"You would think that after all these years, they'd quit wasting our tax dollars," the other replied.

"Enough is enough," her friend agreed.

Azelia bit her tongue and left without the stamps.

TALL LIKE his father and warmhearted like his mother, young Larry left the house early one Saturday morning to keep an appointment with the Milford dog officer. The plan called for him to adopt a puppy from the pound after watching it displayed on television. On TV, the pup looked into the camera with sad eyes an expression that fairly begged, "Please adopt me." Face-to-face at the pound, the dog's tail thumped happily against the side of the kennel and she gave him a smile that melted his heart. He paid the thirty-dollar fee and headed home with his new charge seated proudly on a blanket beside him.

"Name's Maggie," he told his dad.

"What's she look like," Heron asked, as he bent to pat the dog and scratch behind its ears.

"Part Golden Retriever and part German Shepherd. Tan like a Retriever with the dark shading on her head and hind end characteristic of a German Shepherd. Not as big as either, come to think of it, so maybe there's something else in the mix."

"Beautiful. Hello Maggie Heron."

The tail thumped and Maggie licked Heron's hand as her entire body oscillated in happy response to the kind hand stroking her head. She had found a home at last.

THE STORY appeared in Monday's Gazette. Last night, three masked men, armed with a sawed-off shotgun, machete and knife robbed the Immaculate Conception church in Worcester, Massachusetts, tying up the Reverend Edward T. Connors and two young volunteer workers with rope and wire they brought with them. They then forced their captives to lie face down on the floor while removing approximately three thousand dollars from the rectory safe, money gathered in Sunday's collection. The two women, more loosely tied than Connors, freed themselves then the priest shortly after the robbers left the scene.

Immediately after Azelia read the article to Heron, he phoned his friend. "I'm so glad no one was hurt."

"If anything happened to either of the volunteers, I could never live with myself," Connors told him.

"Think you'll recover the money?" Heron asked.

"I doubt it. So far the investigation turned up nothing. Nothing about the robbers stood out, so they're hard to trace. It all happened so fast, we did not get a good look at them. The police think that at least one might belong to our congregation. They seemed to know the how, when, and where with respect to the collection money. God forgive them."

Nothing like this ever happens in these small towns where religious people respect the sanctity of the Catholic Church, another sign of further change that did not bode well.

HERON WAVED goodbye as his brother-in-law drove away from their house then started up the front steps, tripping over a newspaper deposited by the paperboy earlier. He fell forward, smashing a knuckle against the wooden step as he reached out to break his fall. "Damn," he cursed, holding the knuckle to his mouth and tasting blood.

He felt irritable, tired of depending on others for rides, never able to drive to a friend's house or grocery store, or take Azelia for as spin, falling over a damned stupid newspaper he could not see. Then he thought of the many wonderful things he would never do again, what others took for granted. Today a complaint about his affliction finally escaped his lips, though no one stood by to hear it. "Why me?" he asked aloud.

Dusk fell and he mounted the steps under a grainy sky he could not see, crossed the porch landing and reached to open the door and smashed the already sore knuckle on the knob, feeling it tear. "Damn!" When he flung the door open, he called, "Hey Maggie! Where's my girl?"

He heard a thump as she leapt off the sofa in the living room then her claws scrapped on the wooden floor as she ran to greet him. Heron could not see the smile on her face nor the sparkle in her eyes but he did hear her tail whacking the commode in the hall, and that helped lift his spirits.

Both ends of Maggie's body wagged at once; head, ears, and tongue gyrating at one end, tail and butt at the other. Normally she had more of a calming effect on him but not so much today. The deep depression started in the pit of his stomach and clawed upward until it squeezed his heart, the pressure continuing to build inside like a volcano preparing to erupt. "Why me, Maggie?" He bent to stroke her neck and pat her head. "Tell me why?"

He plunked down on the sofa and leaned back resignedly as Maggie flattened obediently at his feet then dragged her body closer.

"Most of my buddies went on to college on the GI bill. They drive BMWs and Mercedes. Ironic isn't it? Cars made in Germany, I mean." He sighed. "Think what might have been..." His voice trailed off. Knowing that continuing down this path would lead nowhere, he did so anyway. "They can afford fancy cars, maids and nannies to care for their kids while I just hope to send mine to college."

Maggie came to a sitting position and sent a lick in the direction of Heron's knee that missed entirely.

"Wouldn't it be wonderful if I could afford nice things for my sweetheart, eh? She deserves a break, wouldn't you say?"

Maggie could stand it no longer. She answered with a yelp.

"You agree then? The best I can do is to not be a burden to her." He paused, then leaned down to give a rub behind the ears. "That's why I go walking around town alone and make my own toast in the morning and ask for no help unless absolutely necessary. How can I ask any more of her? Know what I mean, Maggie?"

Maggie let out a low whine, as though she understood that her master began to cry. No person could detect it, the master's tears falling inside, tears not seen or felt, just a salty taste. But nature gave Maggie special senses. She knew when he felt unhappiness and would do everything in her power to lift his spirits. She drew closer and placed her head on his knee, looking up with soulful eyes, sensing the misery in her master's heart.

Why shouldn't he feel down-hearted? No promises and no future will do that for you. He felt like flotsam tossing on a rough sea, with no bearings, no destination, no dreams or ambitions. His skin felt clammy, sallow, and cold sweat pervaded his body void of substance.

Finally, he stood and made his way to the rear door where he dropped to a sitting position on the back stoop while Maggie slipped past to visit the outdoors and attend to business. The quiet except for a slight breeze that bathed his moist flesh, cool and invigorating calmed him. He drew in a deep breath and tasted the fall air.

Get hold of yourself, damn it! You can't run but you can walk. You can't see but you can feel and hear, and your senses are acute. You don't look beautiful but what the hell – you have Azelia. And she'll always look to you the way she did at age twenty-three. Who else do you know that lucky? So stop feeling sorry for yourself. She deserves better than that. Get on with life.

Heron welcomed the jangle of the phone and rushed to answer it, nearly knocking over a plastic garbage can in the process. He tried to calm down as he moved with the speed of a sighted person to pick it up. The caller identified herself as David's mother, Mrs. Rubenstein. Strange. In all these years he'd never heard from her, and still didn't know her first name. "I just called to see how you're doing, Larry."

"Fine. Just fine, and you?"

"Oh, I'm having one of my mood swings," she said. "I got to thinking about David and it led me to think of you."

A long pause. "Can you tell me something, Larry?"

"If I know the answer, sure."

"Tell me how it happened? I need to know how David died. The Army told me nothing really. I need to know the truth."

He understood. This did not mark the first time someone called to ask how a loved one had died. Many times he listened to the experts on TV talk about closure. There is no such thing, he thought, but knowing the truth might help lessen her pain.

"Would you like to come over to talk about it in person?"

"Oh, that's so nice of you, Larry but I think I'd rather hear it now. I may not be good company... you know."

"Yes." John Sears told him the whole story during a conversation that took place at a recent 87th Battalion reunion. "I was not there but I understand it happened so quickly that none of those killed ever knew. They were relaxed and eating breakfast when a shell came in unexpectedly and hit a tree hanging overhead. The fragments killed them instantly. There was no suffering."

"Are you sure?"

"Positive. They were killed where they sat, some with spoons still held in their hands. They never knew."

A moment of silence, then, "You know…" She stopped to get a grip. "My David was the nineteenth of fifty-five Milford sons to lose his life in World War II. We buried him in the Beth Israel Cemetery in Everett, with full military honors." She paused. "Ironic that he died in France, on July 4, 1944 and two weeks later I received a letter that he wrote to me on June 28 from inside a fox hole. 'Slaughterhouse' is the word he used to describe his surroundings."

"He was a good friend."

"Thank you Larry. I am so sorry for what happened to you. You are very brave. If there is ever anything I can do..."

"I appreciate it, Mrs. Rubenstein. My wife told me how nice you were to her at the train station when we left that day, David and I. She'll always remember you for that."

They said their goodbyes and Larry made a move to attend to Maggie woofing at the back door. He knew he should have for her first name but the opening just didn't arise.

SATURDAY in 1974 opened with a limitless blue sky dabbed with occasional puffs of cotton. Last night, Debbie worked until two a.m., sprucing up the apartment she shared with another student off the Framingham State College campus. Her mom and dad would arrive any minute now and she wanted to impress them. A floral afghan covered the coffee stains on the living room sofa. Although it was her roommate's turn, she scrubbed the bathroom (ugh) and replaced all the soiled linens. No dirty dishes sat in the sink and a fresh fragrant bouquet of flowers graced the table by the entrance.

She had thought of everything, everything, that is, except the bare floors that reverberated with the loudness of a Jamaica kettledrum whenever anyone walked across it, which was the first thing her father noticed as he entered the room. After giving her a hug and asking how she had been, he moved about the room stomping his feet. "You need some carpeting in here," he stated, flatly. "To help keep the noise down"

"I know dad."

"It's so loud when someone walks, you can't hear yourself think," he said later. "I couldn't study if I were you, not with all this racket going on." He stomped a bit more to make his point.

They intended for their visit to last just long enough to drop off a care package consisting of blankets and other essentials, so she did not have to listen to his complaints for very long. After they left, Debbie went to the library and then out to dinner with some friends. When she returned at eight that evening, her roommate, Linda, stood waiting impatiently at the door. "Debbie, you aren't going to believe this."

"What?"

Linda stepped aside and pointed to the living room. "Look. Your parents bought us a new rug."

"They what?"

"Your dad and mom stopped at a rug store in Nobscot, purchased the rug and had it delivered three hours after you waved goodbye to them."

"It fits perfectly. How did they know the size?"

"You got me."

"My mom probably paced it off when we weren't looking. She's good at things like that."

"Your father is something else."

"Oh no, what did he do?"

He called a little while ago. "When I told him I really liked the color, that the colors matches everything in here, he said, 'Of course. I picked it out myself.'"

They both laughed.

He felt so strongly about his children that he always tried to give them the best money could afford. Family came first with Heron, so when Patty married, it meant a great deal to him to sing *Daddy's Little Girl* as the bride and groom danced together. People listened with their hearts and when he finished, Norma leaned toward Azelia and asked, "Isn't he just beautiful?"

The next year, young Larry entered the University of Massachusetts and like his father excelled in sports, playing baseball and basketball for the school.

AZELIA FEELS like a passenger on a train gathering speed as it heads toward a long dark tunnel toward her final stop. She feels at peace and thinks perhaps she should not fight it any longer, just let go. Her eyes close and remain closed until a loud thumping noise jolts her back to her hospital bed and reality.

"Good morning, Mrs. Heron." The neurological department's emissary rolls his portable X-ray machine to a stop beside the bed.

She never saw this guy before. Seems it's always someone different. "Good morning," she tells him. What's the point of looking at these lungs, they aren't getting any better?"

"This'll take only a minute. Just lie still and let me do the rest."

He lifts a corner of the pad that the nurses always leave under her partly for such purposes, and slips a photo plate under her frail body, finally positioning it by giving a good shove with his foot. Satisfied he positioned it properly, he tells her, "Take a deep breath. Now hold it." The machine whirs and he quickly removes the plate from under her. "That's it, Mrs. Heron. You can get back to sleep now."

A moment later he departs.

"Now where was I? Oh, yes the part I hate most to think about."

CHAPTER TWENTY-FOUR

THE SLIDE CONTINUES

THE SHRILL sound of a siren sliced through the morning air like an ax through butter. Eastbound traffic on I-90 southwest of Boston swerved to one side or the other to open a path for the ambulance that cut across three lanes to access the Prudential Center exit ramp. Moments later it pulled up at the emergency entrance to the Jamaica Plain Veterans Hospital where a scene of controlled chaos erupted. The doors to the emergency vehicle popped open and a team dragged a gurney from the rear with dispatch.

Moments later, the emergency room doors parted simultaneously and a doctor rushed forward, took one look at the patient and felt his own pulse quicken. A member of the EMT filled him in on what little he knew as they rushed the patient inside. The doctor made a mental note of his grossly swollen neck and face and the feverish tint of his skin. The patient's responses to questions sounded slurred and barely intelligible. The team could not determine the cause of his suffering, and became horrified upon removing the pair of sunglasses only to find the man's eyes missing.

When they rolled him into the ICU, the doctor listened to his chest then ordered the nurse to hook up an IV for re-hydration and to take his temperature.

"It's 100 and climbing."

"Tell me when it hits 103. I want stat blood work and film of the swollen area," he ordered. It quickly became apparent that the patient suffered from a massive infection that sent him into shock, but the cause remained a mystery.

Fortunately, this hospital boasted a doctor on staff who that same year made major contributions to a new field called craniofacial surgery, the same doctor who would later establish the craniofacial program at Children's Hospital to correct deformities and cleft palates. The emergency room doctor wasted no time seeking to find that particular doctor for a consult. Fortunately, he found the surgeon on duty and available.

"It doesn't look good," he told Dr. Joseph E. Murray, the craniofacial expert who arrived within minutes of receiving the call. "I think we should operate, but frankly I wouldn't dare. I have no idea what's wrong. It could possibly be a pituitary adenoma, requiring transsphenoidal surgery."

Dr. Murray picked up the chart and gasped when he read the name on the patient's chart. When he read the name, Lawrence J. Heron, an immediate sense of déjà vu seized him, and summarily he began to suspect the cause of Heron's new trauma. "It's very likely something else. Do we have X-rays?"

As if on cue, a nurse arrived. "Here are the X-rays."

The doctor scanned the X-rays then said, "Just as I thought. See this?" Murray pointed to a dark spot close to the pituitary gland. "It looks like a piece of metal, shrapnel that's been in his body for years. This man is Larry Heron, an old friend. Our paths seem to cross whenever he's in the most trouble. Let's get him into surgery immediately and see if we can find a way to extract that thing."

"Temperature's 103!"

Murray knew he must act fast. Heron was losing consciousness and he feared his patient might soon lapse into a coma, with consequential brain damage or death.

First, the doctor took steps to drain fluids then formulated a plan. He could open the skull and go in, but on second thought decided that particular procedure too intrusive and inherently the most dangerous. He could try to reach in though the sphenoid sinus, one of the facial air spaces behind the nose but that would not get him close enough and the shrapnel appeared too large to extract through such a small space. He finally decided on the direct transnasal approach, making an incision in the back wall of the nose. Judging from the location on the X-ray, there seems ample room to get in and make the extraction.

After a lengthy and delicate operation, the piece of metal causing the problem landed in a tray with a loud clink. Dr. Murray did all in his power to insure the removal of all possible sources of infection. One more day without attention and Heron would surely have perished. Or would he? This was, after all, Larry Heron, the man who defied death.

As the anesthesia wore off, an old acquaintance returned to pay Heron an unwelcomed visit, intensely visceral pain. But thirty-six major surgeries prepare one for endurance of pain, especially with the help of an old friend, morphine, which enabled him to sleep well that first night. By morning he wanted a real breakfast , except that the doctor denied him anything but liquids. Over the next few days he managed a speedy recovery and on the day of his release, Dr. Murray came by to wish him well.

"We've got to stop meeting like this, doctor," Heron joked.

"What are the odds?" Dr. Murray asked.

Heron extended a hand. "Thanks, doc. You saved my life – again."

"Someone else helped."

"You mean God?"

"Him too but I mean someone else from your past. He's right here."

"Hi, Larry," a familiar voice said.

"Dave? Dave Tredeau?"

The doctor and Tredeau exchanged astonished glances. "How did you know it was me?" Tredeau asked.

"Never forget a voice." Heron said, drawing a breath.

"This is like an episode from This is Your Life," the doctor joked.

Dave played shortstop on the same St. Mary's baseball team with Heron when they both competed against Dr. Murray playing for Milford High.

"You were the guy who threw me out at home plate the day of the slugfest," Dr. Murray laughed.

HERON PREPARED to sing with the accompaniment of the famed Army Band at an outdoor ceremony held to commemorate the addition of a new wing to the West Roxbury Veterans Hospital. Beforehand, he charged Patty with the task of hitting the record button right after his introduction.

She sat ready at a table beside the grandstand trying to stay awake through endless speeches, when suddenly she heard a low drone as a yellow jacket swept past her ear. She held perfectly still, hoping it would fly away but a squadron of wasps then appeared. As she backed her chair from the table, she spotted the attraction – a garbage can parked a few feet away.

The opening speeches by dignitaries ended and Heron stepped to a microphone to begin singing The National Anthem. His voice sounded deep and resonant. This was, in fact, the best performance of his life. Thank heavens he decided to make a tape of it.

The wasps did not seem happy with the music; in fact, it stirred them to action and they began a frantic circling like planes locked in a flight pattern around a busy airport. One buzzed so close that she thought it would land in her ear. She swung at it with her program guide, which only seemed to anger it more, stimulating a change in flight pattern to a closer circle around her head. The wasps began to organize, signaling one another perhaps that the time arrived to sting. She waved the guide with one hand and dragged her chair further away with the other. Locating a shady spot far enough from the garbage can, she unfolded the chair and sat down again.

"Did you get it?" Heron asked before she finally settled down to flip on the recorder.

"Get what," she asked, then abruptly stood. "Oh no. I forgot to turn it on. Oh. No. Did you sing already? Was it good?"

"Just my best performance ever," he said, confidently. "But you're kidding me, right?"

Patty cried, "No Daddy. I didn't turn it on. I'm so sorry."

"What?" he asked, incredulously.

"I became so distracted by a bunch of wasps that I forgot to turn on the recorder. I'm really sorry, Dad."

"Did any bite you?"

"No."

"Well I'm glad of that. No big deal, sweetheart." He encircled her shoulders with an arm. "I'll just have to do better the next time and you too, okay?" He hid his disappointment so well that it almost had her convinced – almost.

"I'm so sorry, Dad."

Heron went off smiling to talk to some dignitary as though nothing at all happened. But Patty Heron would never forget that day. Everyone who stopped to talk to her commented on how wonderful her father sounded. Someone even said, "I wish I'd recorded it. To which Patty replied, "Yeah, me too."

THE DOWNWARD spiral of the Draper Corporation that began in 1967 when Rockwell bought the company, gained added momentum recently. Besides market saturation, Draper looms became too expensive. Three-and-a-half foreign looms cost about the same as one Draper loom. By 1978, the plant closed and the era of the great company town came to a close.

"The one good piece of news I learned today," Heron told his wife. "They plan to let us stop renting our houses and buy them at below market prices just to get them off their hands." After he relayed the price, she replied, "We can't afford not to buy our house for that amount."

The closing of the Draper manufacturing plant forced Heron into retirement. Young Larry would complete college that same year and begin teaching biology at Hopedale High School. The Herons lived off their savings, disability, and social security payments and the money Azelia took in after accepting a job as a librarian at the Bancroft Memorial to supplement their income took some of the pressure off Heron losing his job.

Though patriotism slid into decline, traditions remained strong in Hopedale. On May 28, 1979, Heron played a central role in Memorial Day commemorative services, after which members of the press asked how he felt about the war. "I don't believe in war but when my country calls... well, I love my country. I gave a lot for my country." When asked about the incident that left him so badly damaged, he responded, "We had to return fire and someone had to unload the truck. The decision was mine alone." That concluded what he cared to say about the ephemeral events that changed his life forever.

FATHER EDWARD T. Connors retired from the priesthood in 1979 at age 73. In November of 1980, his successor, Fr. Joseph W. McKiernan, invited him to return to Immaculate Conception to continue his 9th Division "annual pilgrimage" on the occasion of Veterans Day. Retired Chief of Staff General William C. Westmoreland, a long-time admirer of Connors and the man who headed up the 9th Division at the end of World War II, presided over a ceremony in which a committee sealed the names of the division's 250 Vietnam War dead inside a tan brick monument in front of the church.

At a similar ceremony in 1966, Westmoreland presided over the sealing of microfilm containing the 4,581 names of 9th Division soldiers killed in World War II. Major General Louis A. Craig, the commanding general of the division during most of the Second World War showed up to participate this year. Members of the Worcester County Chapter, Vietnam Combat Veterans, and Combined Allied Forces all took part in a parade that kicked-off at the Worcester Fire Department on Grove Street and ended at the church overlooking Gold Star Boulevard. Ceremonies ended with the annual division banquet held later that day at the Sheraton-Lincoln Inn.

Fr. Connors, Gen. Westmoreland, and Fr. McKiernan.
Courtesy Bishop George E. Rueger, Worcester Diocese.

During Mass, Fr. Connors hailed all veterans "who laid down their lives in the cause of peace." To the Vietnam Veterans, he added, "Your lives have been far more difficult than ours because we were allowed to win our war."

"Westy," as those who knew him best referred to General Westmoreland, gave the final address. The Supreme Commander of Allied Forces in Vietnam for more than four years, he praised veterans of the conflict saying that U.S. troops entered Vietnam "to stop the ruthless Communist regime of Hanoi from overrunning South Vietnam." He added, "Now all Americans regard that as a noble cause, especially when they see events shaping up in Indochina today. Nobody wants to go to that part of the world again. They all want to get out."

He termed American withdrawal from Vietnam a fault not of the soldiers who fought there. "They did the job they were sent to do. It was not our soldiers who failed. As in World War II, the men who fought in Vietnam supported a principle that has always characterized Americans: peace."

At the close of ceremonies, Heron sang the National Anthem then from a church balcony bandleader, Harry L. Bullens played Taps, the bugle call for the military dead, while a second bugle player, hidden in a distant location, played a second, haunting Taps echo, sending chills tracing down spines and drew tears from all eyes.

At the completion of ceremonies, Fr. Connors introduced the Herons to Fr. McKiernan who told them, "Last June, in a surprise celebration, the Catechetical Center at Immaculate Conception was renamed the *Reverend Edward T. Connors Center.*" The word Catechetical means a place where different religious classes are taught and in this case, the Catechetical Center occupied one entire hall of the church.

"That is so nice," Azelia said. "He certainly deserves it."

"Besides, no one could pronounce 'Catechetical,'" the priest joked. "At the celebration, Father Connors said that though embarrassed, he was grateful for the tribute. The parishioners showed how much they loved him and referred to him as 'mister-one-in-a-million,' and his soldiers continued calling him the Godfather of the 9th Division."

"He is one in a million."

"A hard act to follow," McKiernan concluded.

IN THE SUMMER of '84, the National Veterans Wheelchair Games took place in the town of Brockton, Massachusetts, and what better tribute than for Lawrence J. Heron to open the ceremony with the National Anthem accompanied by the 18th U.S. Army Band from Ft. Devens?

Cardinal Bernard F. Law gave the Invocation.

Then World Middleweight Boxing Champion, Marvelous Marvin Haggler, appeared with George Lang, Congressional Medal of Honor winner, for the lighting of the torch. Following the ceremonies, the owner of Ben's Tavern offered Heron and his friends, Guido, Ed, and Pete Ferrelli, free coupons for food and drinks at his tavern.

Located off Bellevue near Main Street, Ben's sounded noisier than usual on "Buck-A-Burger" night. Azelia did not accompany Heron because of a previous commitment to attend a Bingo tournament in Milford with her sister, Olga.

Above the cacophony of laughter, music, conversation, and Red Sox play-by-play, from the bar came the gravelly voice of a heavy-set man. Bubba, as his friends called him, fell into a foul mood because the Red Sox just lost to the Yankees.

Heron sat at a table near the piano at the back of the room where the piano player started playing while Heron sang *Heart of My Heart* and most patrons joined in. After *Four Leaf Clover* ended, someone called for Heron to sing *Forty Shades of Green.*

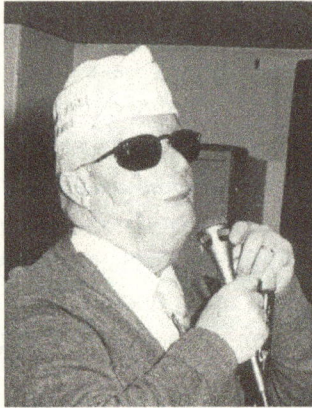

A popular figure at functions, especially patriotic events.
Courtesy of Larry Heron, Jr.

As he began singing in his soft mellow voice, the noise in the room abated. Everyone listened quietly except Bubba. "Hey, bartender, another bourbon!" he bellowed. He swallowed the bourbon, savoring its sharp descent then followed with a swig of beer from a frosted mug. Wiping his mouth on the sleeve of his green and brown-checkered polyester, he let out an exaggerated burp and plunked his empty beer mug down noisily. "Gas 'er up," he said. "And how about some quiet so's we can hear the TV?"

The bartender frowned and said, "Hold it down, please. You're disturbing the customers."

"You mean the blind guy? He's disturbing me!" Bubba said loud enough for everyone to hear.

"I am asking you politely, Buddy." The bartender said again.

Bubba slid his glass forward. "Just pour me another."

The bartender complied.

Upon request, Heron began singing Danny Boy.

"When's the blind guy going to give it a rest?" Bubba shouted as his friends laughed encouragingly.

Heron raised his voice and the patrons threw disgusted glances Bubba's way. When the singing ended to applause, a woman at a table near the bar leaned toward Bubba and said, "Show some respect. That man's a war hero."

"War he-ro? Hey, Audie Murphy," Bubba shouted. "How many babies you kill?" His cronies slapped their thighs. Thus encouraged he continued, "So what's with the blind guy anyway? How'd he get his face messed up?"

The woman at the table rose in protest. "That's an awful thing to say."

"Pipe down," Bubba blurted.

Pete Ferrelli, built like a linebacker, pushed his chair back from Heron's table. "That's it!" he said. "I'll shut him up."

"No!" Heron extended a hand in protest. "Let the blind guy do it." As he tapped his way toward the bar with his cane, the room became pin-drop still.

"The blind guy leaving us?" Bubba asked his friends.

His voice served as a beacon.

Heron homed in.

Suddenly, Bubba grew still. He could smell his own fear as the blind guy drew closer to him with the bearing of a soldier, head held high. The nearer he drew, the more he could swear the blind guy could see past his eerie glasses. The scarred face looked ashen and the mouth curled into a crooked smile. He began to wonder if perhaps he had gone a bit too far.

"You should apologize to the lady," Heron said flatly, his nose inches from Bubba's. Though somewhat unnerved, Bubba nonetheless felt safe. After all, this man was totally blind and Bubba's friends surrounded him. Instead of apologizing, he said, "I don't see no lady."

"You have three seconds," Heron said.

"You'd better go back to your buddies."

The blind guy appeared confused. His hands groped for the bar. Bubba suddenly felt more in control. "Think 'cause you're blind I gotta take your crap?" He jabbed Heron sharply in the chest with an index finger to punctuate his claim.

Heron counted, "Three, two, one."

Pete, Ed, and Guido rose simultaneously to their feet in awe as

Bubba rose above the crowd like Tinkerbell. The big man's eyes bulged out of his head. A split second later he became airborne, flying over the bar, clearing the glasses and bottles to come crashing down on the floorboards behind. When he opened his eyes the bartender stood smiling down at him.

The blind guy leaned over the bar and appeared to look down at him as well, "How's that for a blind guy?" he asked.

A cheer erupted as Ed took Heron by the arm and led him triumphantly back to their table where the singing commenced. Now everyone joined in as Bubba and his friends quickly settled up and departed silently through the front door.

A patron leaned forward and asked the woman who spoke up earlier, "What the hell just happened? Who is the blind er...gentleman?"

"A living example of what makes this country so great," she answered.

CHAPTER TWENTY-FIVE

FR. CONNORS LAST FAREWELL

TUESDAY, January 28, 1986, opened bold and blusterous and continued to worsen as the day unfolded. When Father Thomas O'Malley stepped from his car in the visitor's lot outside St. Francis Home, he heard the wind groan like an organ playing off-key. As he trudged across the crusty snow, he ignored the freezing temperatures to inhale the freshness of the cold night air. The call came in the middle of the night, as often proved the case. This time the summons directed him to the bedside of his dearest friend, the ailing Father Edward T. Connors.

Connors long since prepared for death and did not fear it, his faith that strong. If he could no longer go about helping others, then perhaps it was time to go. Nevertheless, when he thought back to the life he lived, sadness overtook him, for there remained so much more to accomplish before leaving this world forever.

Fr. O'Malley leaned forward to look down upon this once vibrant priest and dear friend lying flat on his back for the first time since he'd known him. His face looked drawn and eyes hollow. His arms appeared frail and his face had taken on a deathly pallor.

But the kindly smile remained.

Never before did he hear Fr. Connors speak about his past, but tonight was different with time running out, so he talked of old times and revealed a few secrets.

Connors' love of sports and the kids he helped shape into responsible men, meant more to him than simply winning games. He believed a great deal could be learned on playing fields, and that every young person could benefit from involvement in sports. Every kid deserved the opportunity to achieve a "personal best."

But it wasn't simply healthy bodies and minds that concerned him. He felt convinced that playing sports builds character, that it teaches respect, trustworthiness, honesty, responsibility, fairness, caring, and good citizenship. He looked up at Fr. O'Malley, and asked through blurred vision, "Remember the Sullivan brothers?"

"Three of them. It surprised me when they all went to college and turned out so well, but as kids, they were a handful."

"They certainly were." Connors eyes looked up at the ceiling. "I may have had a hand in how they turned out, at least I hope so."

"I'll bet you did. Tell me about it."

"Well, you see this one time the boys acted pretty rowdy following a basketball game. In those days, we had no shower rooms or lockers, just old Flanagan's barn rigged with a shower."

"I remember it well."

"I heard a commotion that day, and when I went to investigate found the boys teasing poor old Randy Cowan, the janitor who cleaned up after them. Cowan was mentally impaired but it never stopped him from working hard and it didn't mean he wasn't sensitive. The boys didn't know that or didn't seem to care. They were jeering and snapping their towels at him, teasing the heck out of the poor fellow. Cowan looked really upset, like he was crying. That really got to me."

"So I lined the boys up in that barn, all of them. Then I went right down the line from left to right, slapping each and every one of them across the face, including Bobby Sullivan. By the time I finished, my hand felt plenty sore. Then I sent them home thinking about what they had done."

"The next day I received a call from Mrs. Sullivan. She said, "Robert tells me you slapped his face yesterday.""

"'Yes I did,' I replied."

"Good. I just called to tell you that if he ever does anything like that again, I want you go right ahead and whack some more sense into him. Keep him straight."

Fr. O'Malley flashed an amused smile.

Connors grinned up at him. "Can you imagine what that would

cost me today?"

O'Malley fought to hold back a flood of tears. "Times have changed a great deal in our lifetime," he said. "And I'm not sure it's all for the better." Then he added, "Someone should write a book about you, Father Connors. You've helped so many confused kids straighten out their lives and go on to become responsible adults. And the way you volunteered to serve alongside them in battle..."

"I have often prayed that when my work here is done, which it appears could be at any minute now, He will tell me, 'Well done, Edward, well done.'"

Fr. O'Malley stayed by his side, talking to him while Connors' eyes involuntarily closed. O'Malley refused to leave his side until the end, which he suspected not long in coming. And when the end finally arrived, when Fr. Connors left this earth and the people behind, those he loved so dearly, O'Malley sat a long time remembering his dearest friend.

The next day, he sat beside Fr. George Rueger discussing some of what passed in Connors' time. Fr. O'Malley said, "When the boys would not come to confessional, he would bring the confessional to them – in the rear of his beach wagon." Remember the vehicle with the wooden siding that gained overwhelming popularity throughout New England during that period of time? "The kids would sit on the tail of the wagon while Father Connors listened through a drop-down canvas sheet. He even took confessions behind trees and in open fields."

"Sports played a major role throughout his life," Rueger said. "But I think the place he felt he did the most good was in the war as a chaplain."

"Yes. You know he took great pleasure poking fun at himself. He used to say that he was the man who never got the bird of a colonel or the red of a monsignor, and then Westy conferred upon him the honorary promotion to *Green Colonel.*"

"I remember," Rueger laughed, though his ears grew teary. "He was very proud and quite moved by that. The general thought quite highly of him, didn't he?"

"In August of '66, Westy sent him a wire from Saigon asking his old and valued friend to accept an award on the general's behalf at ceremonies held in Miami Beach. The general was off fighting a war and could not attend, so Fr. Connors wired back that he'd be honored."

"Without a doubt," Rueger said, "he was the most popular and widely known priest in the diocese. He'll be sorely missed."

ON WEDNESDAY, April 16, 1986, Massachusetts Governor Michael
Dukakis waited on the steps of the oldest building on Beacon Hill for the
arrival of a state hero, Lawrence J. Heron. The State House, completed in
1798, sat atop a parcel of land once owned by John Hancock, the state's
first elected governor and signer of the Declaration of Independence.
Native-born architect Charles Bulfinch designed the State House that
overlooks Boston Common and the Back Bay with its copper dome coated
with 23-karat gold, completing the project in 1798.

The governor, local representatives, and DAV officers gathered to
give special recognition to Heron, the now legendary hero. At the conclusion of ceremonies, the governor accepted a POW-MIA cap and key chain
presented to him by Heron in memory of those from Massachusetts still
missing in action.

The amiable governor appeared genuinely honored to meet Heron,
asking about his war experiences, but Heron steered the subject away
and the two men began swapping views on subjects ranging from raising
children to recent sports activity. When Heron mentioned his wife's name,
Massachusetts State Representative Marie J. Parente piped up, "She was
named after the flower, but her parents had just come over from the old
country and misspelled it."

Governor Dukakis smiled. "After seeing her picture, I believe the
flower got its name from her."

After posing for photographs, Dukakis invited his honored guest
on a guided tour, and Heron found the central chamber of the House of
Representatives most impressive, with its massiveness clearly echoing in
his sensitive ears.

ON JUNE 6, 1990, a week after Memorial Day, Representative Parente referred to Heron in a monthly State House Report published in the
Milford Daily News. *Children could learn a lot about the importance of
saluting the flag from Lawrence J. Heron, for whom the Milford Disabled
Post is named. Each time the American Flag and its bearer approaches
his "parade spot" (guided by a whisper from a friend) Larry's quick salute
gives viewers a glimpse of what he "sees" in its beauty and significance. It
is because of sacrifices like those of Larry and his counterparts, that Old
Glory still waves and stirs the hearts of his Blackstone Valley countrymen.*

Larry, on behalf of children (and all of us), thanks for your lasting and selfless devotion to our well-being.

That about summed up the feelings of the people in southeastern Massachusetts for Larry Heron, an icon, a living symbol of proud veterans everywhere, one who loved his country enough to surrender his most precious gifts to preserve it.

Heron fishing with flies he made. Also loved to swim and play golf.
Courtesy of Larry Heron, Jr.

"OH MY!" Azelia exclaimed.

"What?" Heron asked.

"Dr. Murray made the headlines again." Her voice trembled with pride as she read from the October 9, 1990, issue of the Boston Globe. "*Joseph E. Murray, 71, of Wellesley, professor emeritus of plastic surgery at Brigham and Women's Hospital is the Recipient of the Nobel Prize in Medicine for his discoveries concerning organ and cell transplantation in the treatment of human disease.*"

"The award was given for performing the world's first successful organ transplant in 1954, and proving it possible to transplant organs between non-identical relatives and from the deceased to the living. They estimate that each year 20,000 people receive a new lease on life because of him."

"Here's another article in today's Worcester Telegram and Gazette where he mentions how he came upon you at Valley Forge and didn't recognize you until reading your name on the chart. '*Working with burn victims at Valley Forge General Hospital during World War II inspired the doctor to steer his medical career towards tissue and organ transplantation.*'"

"The Nobel Prize? Wow! More than the money, it's the prestige. He's climbed the Mount Everest of prestige," Heron said, with admiration.

"Funny you should say that. It says here that he once scaled the Mattahorn."

Heron remained quite active as he aged.
Courtesy of Larry Heron, Jr.

Chapter Twenty-Six

Died on the Fourth of July

Upon finishing grocery shopping in Milford on Thursday, the Herons returned home to find young Larry waiting to greet them at the door. "Some guy from Boston called about an hour ago, Dad. Didn't leave his name but said he'd call back later."

"Wonder what he wanted?" Azelia said, as she plunked two bags of groceries down on the kitchen counter. Heron entered with two more bags. "If it was important, he'll call again," she said, taking the bags from Heron.

The phone rang again at 7:10 p.m. and the man on the other end identified himself as the artistic administrator for the Boston Pops, calling on behalf of Keith Lockhart, conductor of America's Orchestra. "Some of us sense a resurgence of patriotism, something the country sorely needs. I've seen Mr. Heron's picture in the paper and heard him sing the National Anthem at Memorial Day services last May. No one stirs patriotism more than Mr. Heron. We would be honored if we could persuade him to open this year's 4th of July ceremonies with The National Anthem."

Words suddenly formed a traffic jam in Azelia's throat. "Why, er, ah, just a moment. He's right here."

Azelia watched his face light up. "At the Esplanade?" he responded. "That would be..." He hesitated. "Quite an honor," Heron said with a smile.

When he hung up the phone, Azelia threw her arms around him.

"I'm so proud of you," she said as she gave him a firm hug. She knew how dearly he loved his country and how honored he must have felt. No greater patriot ever walked the face of the planet. "I can't wait to tell our friends."

While she left him sitting in the kitchen a few moments to go upstairs to change, Heron headed straight to the back door, opened it, and stood breathing in the raw freshness of the rain. Suddenly an exhilaration swept over him. This represented the first time he felt this alive since the birth of his children. When Azelia came back down, she heard him humming softly.

Azelia, his eyes at every event, made all the difference in Heron's life.
Courtesy of Larry Heron, Jr.

HE SEEMED to enjoy life about as much as anyone could under the circumstances these days, and especially happy about his family, with all three daughters married and several new grandchildren, all delightfully well-adjusted, educated, and productive in society. Not surprisingly, Carol's son, Matt, became a sport's all-star in high school and several colleges pursued him. Although Heron could not see him play, he enjoyed chatting with Matt about each game, and could not have been prouder of his grandson.

Life seemed to run so smoothly that it came as a great shock to Azelia when a week later Heron suddenly collapsed on the kitchen floor. Admitted to the Milford Hospital, he was subsequently diagnosed as having suffered a major stroke.

TWO DAYS LATER, Azelia answered the phone and Heron spoke before she could answer. "Come get me out of here," he begged.

"What's wrong?"

"Just come and take me home. Please. I don't want to stay here another minute."

During the drive home, Azelia asked, "What happened?"

"Nothing's like it used to be," he said flatly.

"Did somebody do something? What's wrong?"

Slowly, she drew it out of him. A young nurse yelled at him repeatedly and today she called him both a "pain in the neck" and a "spoiled brat."

"Did you do anything to deserve that kind of language?"

"Nothing on purpose." But later Heron confessed he felt too embarrassed to use the bedpan, so he climbed out of bed to find his own way to the restroom. Unknown to him, the nurse had moved a table and chair around while he napped and he tripped over the repositioned chair. Then the nurse berated him, accusing him of wanting special privileges and issued strict orders as to how he should behave.

"She had no right to talk to you that way!"

Azelia vowed to keep him home, promising she'd look after him from then on. Times had changed and so did attitudes. In the old days her husband would have been treated differently, with great respect verses humiliation.

Heron grew a uneasy after that, sensing a weakness within never experienced through all the years past, and he now found himself back in bed where it all started fifty years ago, on his back. Was he waiting to die? Where did the time go? If only he could have some of it back. He was not ready to leave a world that held so many people he loved and cared about, and he had more to experience, like singing at the Esplanade, attending Matt's football games, meeting the next grandchild to come along.

When the next series of strokes hit, he refused hospitalization and Azelia did not push for him to return. The doctor concurred that in his present condition, it might serve him better to remain at home. He required constant attention now and Azelia could not leave his side unless someone came to take her place. This continued for several weeks during which he ate very little and couldn't seem to hold anything in his stomach. When

Azelia wasn't caring for him, she was cleaning up after him.

One day, right after giving him a sponge bath in his bed, Azelia, who did not feel so well herself rose from beside the bed. "I've got to go downstairs and prepare supper," she said while heading for the door

"Go ahead, I'll look after him," Debbie told her mother.

When Azelia moved out of earshot, Heron asked, "Who was that woman?"

"That was Mom." When he failed to show signs of recognition, she added, "Azelia – your wife."

"Oh."

Moments later, he drifted off to sleep. Debbie remained by his side, looking down at him with tears in her eyes. He never looked this vulnerable before, not in all the years she knew him. Heron was her rock, someone not kept down for long. She wanted do much to help but could think of nothing to do but pray for him.

A lifetime of unconditional love and devotion.
Courtesy of Larry Heron, Jr.

SUDDENLY HE was eighteen again, wearing saddle shoes and khakis with his hair cut short. Three white stripes graced his blue and gray letterman jacket. Lawrence J, Heron was captain of St. Mary's football team again, a cocky one at that, with always a gleam in his eyes. "Today I'm going to score two touchdowns," he told her, and her eyes smiled back at him as he added, "Just for you."

The game played again in his mind. He scored the winning touchdown and the crowd cheered, and that night, he strutted up Mendon Hill like many other nights, cold air cutting through his jacket like a knife. He stuffed his hands into his jacket pockets then withdrew them, finding he could move faster with his hands swinging by his sides. And there she was, practicing field hockey on the front lawn, as lovely as a summer's breeze.

She stopped when she saw him coming and rushed towards him with arms outstretched. Then she vanished into thin air like a ghostly apparition. And he felt cold, very cold. The bright light ahead appeared inviting. Maybe if he moved closer, he could get warm. He headed for the light.

Everyone in the household heard Debbie's screams and came running. The doctor wrote cause of death as subarachnoid (brain) hemorrhage.

Azelia was stunned. The absence of his company, the silencing of his voice, the void that no other person could ever hope to fill opened like a giant sinkhole to abruptly suck away her life. He will never sing with the Pops but he made it past July 4th. Ironically, perhaps mercifully, the end came on July 7, 1995.

Letters of condolence poured in from friends, relatives and people who had served with Heron. Messages came from people Azelia never met, and from far away. One typical letter arrived from nearby Hopkinton, Massachusetts with an old news clipping enclosed.

The writer referred to Heron as an unforgettable and formidable competitor. *Though we only met on the playing field, I always kept up on Larry and admired the courage and determination you both demonstrated in forging a solid family life. As a fellow veteran, I appreciated how he faced ordeals and daily challenges.*

Larry was an outstanding athlete in every way. He had speed, durability, and was tough to bring down. Our games with St. Mary's were intense and fiercely competitive. Larry was the difference. He was a clean player, aggressive, and I suspect had an innate determination to succeed at whatever he attempted. I also recall that he was very handsome – and I suspect very popular. It seems he faced life as he faced every contest in which he participated. May he now rest in peace. My sincere regards to you and your family.

THE HIGHLIGHT of the funeral was a eulogy delivered by young Larry who expressed in glowing terms the tremendous pride, love, and admiration his family felt for his father. His children had learned so many lessons from a father who overcame every obstacle thrown in his path, and a mother who stood gallantly beside him every perilous step of the way. They could only hope that perhaps now their father could see what his children looked like.

The funeral ended with the melancholy strains of taps, its restful vibrations lingering in the hearts and minds of the hundreds of people gathered there, long after the last note gently stirred the air.

A single tear made its way down Azelia's cheek. Last night and this morning, she cried her eyes out. Her beauty remained striking despite the gray threads in her hair. The years of worry and stress did age her, but gracefully like a fine wine. She at once became the grand dame, incongruous amongst peers, as gracious and still as wise, the matriarch of the Heron family. She reigned magnificent, regal on the outside, but within, dying, a little piece at a time. Goodbye my love. Someday soon I will return to sleep with you - forever.

The following November marked the first time in nearly half a century that Heron did not appear as the central figure in a patriotic event. The name Lawrence J. Heron continued to resonate in newspaper articles and at public observances such as Veterans Day and the 4th of July, Larry Heron would be remembered as "one of the most courageous of America's World War II heroes."

People remembered his athletic achievements, the long suffering, painful operations, blindness, singing, and his deep religious faith. They talked about the characteristics that inspired the nation's Disabled American Veterans to honor him as the only living veteran to have a chapter named for him.

At the same time, Azelia received accolades for her courage and strength. On November 5, 1995, Daniel P. Reilly, the Bishop of Worcester, conducted a Pontifical Mass in her honor and presented her with a special award at the Sacred Heart Church of Hopedale. Pastor Raymond Goodwin said, "It is indeed a precious moment because she is a person who over these many years has given her heart to the Lord, to her family and friends, and especially to her husband, Larry. Azelia remains a living example of God's gift to His church, and in a special way, to our Sacred Heart family." The news media reported on the event and it meant a great deal to Azelia because it helped keep the memory of Heron's sacrifices alive.

Two weeks after receiving her award, Azelia's sister, Olga, passed away. The stress of losing yet another loved one left Azelia weak and drained, but did not prevent her from driving to the Marshfield cottage for a pre-arranged meeting with her children. As she stepped from the car, she collapsed onto the driveway unable to walk. When her son suggested he take her to a hospital, she insisted on climbing into bed instead, believing that she merely suffered from exhaustion.

Her throat felt raw and she exhibited flu-like symptoms, so she asked him for some nasal spray. He rushed to a drug store and when she applied the spray, immediately began to scream and cough up blood in large amounts. With blood pouring from her mouth and nose, family members crowded into two cars to take her to the Jordan Hospital in Plymouth. There doctors took one look at the blood-soaked towels Patty held to her nose and the bloodstained front of her dress and estimated she had lost roughly three pints.

Patty, Azelia, Carol and Larry, Jr. *Courtesy of Larry Heron, Jr.*

The bleeding continued, so the small Jordan Hospital with its limited facilities transferred her by ambulance, still bleeding, to the Framingham Union Hospital where doctors once again failed to diagnose her condition. Since the bleeding refused to stop and she was having difficulty breathing, a doctor stuffed her nostrils with gauze and transported her to the Milford Hospital.

"Perhaps it's lupus," one doctor surmised. "Perhaps her carotid artery is blocked," ventured another. Two years would pass before her condition was diagnosed as pulmonary fibrosis. One of her many checkups uncovered a fracture of the spine from osteoporosis and she was placed on heavy doses of steroids.

Pulmonary fibrosis is a debilitating lung disease that hits heavy smokers. Since Azelia never smoked, she attributed the disease to the reddish-brown particles of dust billowing from Draper smokestacks to settle on rooftops, cars, and outdoor furniture like fine snowflakes over the years. Several people who lived close to the factory developed respiratory problems, many in the same families.

Then on Saturday, May 13, 2000, young Larry came home carrying a large bouquet for his mom. Tomorrow was Mother's Day. He found her in the living room tethered to an oxygen tank via twenty-five feet of plastic tubing that snaked ubiquitously behind her wherever she went. Today she looked so frail and unhealthy that Larry grew more concerned than usual. A constant diet of pure oxygen rendered her cheeks aflame and face pale, yet her bright eyes remained as cogent as ever and she sat there well-dressed as if to go calling.

"Ohh! What beautiful flowers," she told Larry. "You're so thoughtful. Just like your father."

She pondered a beat then added, "They've forgotten him, haven't they?"

"Of course not," he said, unconvincingly. "They'll never forget him." The Herons celebrated Thanksgiving and Christmas at Azelia's that year, since she couldn't travel anywhere without a supply of oxygen. Larry continued to live at home so he could look after her.

JUST AS THE moon reaches the far right side of her hospital window, an alarm sounds and a nurse moves swiftly down the hall to Room 201. The nurse notes Azelia's vital signs registering erratic while her gasps seem to have intensified. She summons a doctor who administers medications knowing that it is not her life that the stimulants prolong, but a slow death instead. He orders a new array of tests then leaves to check on other patients.

Azelia sees Heron's face now, so handsome. His bright blue eyes smiling at her, beckoning, his arms open wide. Look! His face! No scars. How handsome! Why, he looks just as he did at age twenty-three. She runs to him and their lips meet. He holds her close and she smiles through her tears. "Hold me my darling, just hold me!"

THAT MORNING, July 4th 2000, the family received notification that Azelia Heron passed away peacefully in her sleep – which was not true at all. She hadn't slept a wink. She lay awake throughout the night, following the moon on its journey, right to the end, hoping that from somewhere up there, he was looking at it too.

She died on America's most patriotic day, the Fourth of July, almost five years to the day her beloved husband passed on. It seemed their lives meshed like clock gears from the start, always in unison, always keeping good time, reunited at last.

The crowd that gathered at the Heron residence on Christmas Day 2000 had thinned considerably but seemed no less joyful than in prior years. The only guests from the old crowd included Norma, Amelia, and a few other close relatives.

There were many grandchildren now, ranging in age from nine to twenty-one. Carol's son Matt would play football for Maryland next season and eventually wear a Miami Dolphin's uniform. After Christmas dinner, the family gathered in the kitchen where the conversation drifted back in time.

"And do you know, one night he promised she would never see him drink beer again?" Carol asked. "Well, she never did and he never came home with liquor on his breath after that."

"So many amazing stories," Debbie said. "Like the poem Dad wrote for her."

"What poem?" Patty asked.

Carol described the breakup that occurred because Azelia failed to show at John Heron's funeral then read the words to the song from a tattered piece of paper she scribbled out that her mother once recited for her. When she finished reading, she folded it and returned it to her purse. The room fell quiet.

Patty asked, "Remember the time he sang Rudolph the Red-nosed Reindeer at Johnny Milan's funeral?"

"Not everyone's heard it," Carol said.

"Every Christmas, Johnny Milan would come by with his wife, Ester and before they'd leave, he'd sing Rudolph the Red-nosed Reindeer. So when he died, Ester asked Dad if he would sing it at Johnny's funeral. Dad thought it might be in bad taste but she insisted, after all, she told him, 'It was his favorite song.'"

"Well you had to be there. When he started singing, people thought he'd lost his mind. Some laughed out loud, you know, nervous laughter. Others became outraged. It took months for Dad to live it down."

Debbie asked, "Did Mom ever tell you the story about Grandma Livia and how she came to this country?"

Heads shook negatively.

"A Cinderella story, and it's true." She told how the mayor of a small town outside Pisa approached the widowed Livia with a marriage proposal. He told her, "You're a good mother in need of income and I want a child. Marry me and I'll care for you and your children, and we can have a child together."

He soon got his wish. The mayor lavished everything he had on the child born to them, while he condemned stepdaughter Livia to a life of picking grapes in the fields. Along came Grandpa Noferi, a Draper Corporation salesman who spoke fluent Italian. He took one look at Livia toiling in the vineyards and was smitten. Debbie closed with, "They married, moved to Hopedale, and the rest is history."

"I never heard that one before," Larry said. Now that his turn arrived to tell a story about his parents, the women began filing out of the room, knowing he would talk about his father's football games, but the men and his three sisters remained spellbound. When he finished, Patty, who had been staring pensively ahead, said, "Dad was quite a sports star, wasn't he?"

"God, if he had only gone to Notre Dame," Carol said, wistfully, "who knows what might have been?"

"He was the best," Larry said, his eyes glistening. "Mom was too. They both were."

EPILOGUE

ON FEBRUARY 23, 2001, John Sears parked his car outside the Plymouth post office. The front license plate on his vehicle read "Camel Green." In his hand he held a package he intended to mail to Keith and Mary Ostrum in Erie, Pennsylvania, regarding the 87th Battalion's upcoming reunion. He made it as far as the curb in front of the post office where he collapsed on the sidewalk and died. John, a World War II historian kept meticulous volumes of photos, data and records of everything military he could obtain, especially if it had to do with the 87th. The vanity license plate on his car even read CAMEL GREEN. Before he passed away, he donated volumes to the Plymouth Public Library.

The 87th held its next reunion in Gettysburg on September 16-17, 2002. Like many of the ninety surviving members of the 87th, Roger Burt and Angelo Bastoni chose not to attend due to the difficulty of travel to Gettysburg, which requires several modes of transportation to reach. Both Angelo and Roger continued to resided as neighbors in Florida for some time before Angelo moved back north.

A SMALL group of surgeons came together at a symposium in Baltimore in June 2003 to place into permanent record their collective experiences while working with World War II Veterans at Valley Forge General Hospital. Dr. Joseph Murray and Dr. Bradford Cannon, two great American doctors attended.

Dr. Cannon lived into his nineties, passing away on December 20, 2005. His daughter, Sarah Cannon Holden, reported his group performed more than 15,000 operations. In the 1950s, Dr. Cannon served as consultant for the U.S. Atomic Energy Commission and personally visited the Marshall Islands to study effects on the population of radioactivity from atomic testing. He also served as Professor of surgery at Harvard Medical, President of the New England Society of Plastic Surgeons, President of the American Association of Plastic Surgeons, and included among his many achievements, the establishment of Mass General's first plastic surgery residency.

According to this great doctor, rebuilding the wounded at Valley Forge General Hospital, where he continued his work until the ebb of World War II wounded ceased to flow, stood out as his life's crowning achievement.

BEYOND RECOGNITION highlights the lives of several great men and women, notably among them, Dr. Joseph E. Murray, who in 1954 performed the world's first successful renal transplant between the identical Herrick twins at Boston's Peter Bent Brigham Hospital. Medical history records him as the doctor who performed the world's first successful organ transplant. Murray, who served as Chief Plastic Surgeon at Children's Hospital Boston from 1972-1985, "retired" from Harvard University as Professor of Surgery Emeritus in 1986, but remained very active in the field of medicine. His never-ending curiosity led him over the years to work in the fields of transplantation biology, the use of immunosuppressive agents, and the mechanisms of rejection. Discoveries in the 1960s of anti-rejection drugs azathioprine, Imuran, prednisone, allowed Murray and doctors like him throughout the world to extend their technology to include transplants from unrelated donors.

Dr. Murray accepted numerous awards; he published books and articles, and continues to make contributions to the field of medicine. Named Honorary Chairman of the U.S. Transplant Games in the summer of 2004, he appeared as a special guest speaker, as he had so many times throughout his life.

An excerpt from his autobiography speaks volumes about this great American: "We have been blessed in our lives beyond my wildest dreams. My only wish would be to have ten more lives to live on this planet. If that were possible, I'd spend one lifetime each in embryology, genetics, physics, astronomy and geology. The other lifetimes would be as a pianist, backwoodsman, tennis player, or writer for the National Geographic. If anyone has bothered to read this far, you would note that I still have one future lifetime unaccounted for. That is because I'd like to keep open the option for another lifetime as a surgeon-scientist."

Perhaps more than at any time in the glorious history of our nation do we need more heroes like those depicted throughout this book, such as the doctors, nurses, clergy, men and women who serve in our military, sacrificing the pleasures of the good life most of us enjoy and so often take for granted.

BIBLIOGRAPHY

Alexander, Bevin. *How Hitler Could Have Won World War II*. New York: Crown Publishers, 2000.

Ambrose, Stephen E. *D-Day*. New York: Touchstone, 1995.

Ambrose, Stephen E. *The Supreme Commander*. (2nd ed.) Jackson: University of Mississippi, 1999.

Ambrose, Stephen E. *The Victors*. New York: Touchstone, 1999.

American Printing House for the Blind. *Father Thomas Carroll*. Key word: Avon Old Farms + World War II. [Online] http://www.aph.org/hall_fame/carroll_bio.html, 2002.

Astor, Gerald. *The Greatest War*. Novato: Presidio, 1999.

Benison, Saul, and Barger, A. Clifford, and Wolfe, Elin L. *Walter B. Cannon: The Life and Times of a Young Scientist*. Cambridge: Belknap Harvard, 1987.

Bradford, Dr. Vance A. *Burns in Atomic Disaster*. Oklahoma: State Medical Journal, 1961.

Bradford, Dr. Vance A. (1944) *Heron's Medical Records*. Doctor's Report while Heron was a patient at #158 General Hospital, Salisbury.

Breuer, William B. *Bloody Clash at Sadzot*. St. Louis: Zeus, 1981.

Brookesmith, Peter. *Sniper.* New York: St. Martin's, 2000.

Cannon, Dr. Bradford. *Heron's Surgeries.* Valley Forge General Hospital Archives 1944-46. Phoenixville, 2000.

Carroll, Andrew. *War Letters*. New York: Scribner, 2001.

Carroll, Rev. Thomas J. *Blindness: What it is, What it does, and How to Live with it.* Canada: Little, Brown & Company Limited, 1961.

Chase, William H. *Five Generations of Loom Builders: A History of Draper Corporation.* Hopedale: Draper Press, 1950.

Connors, Father Edward T. *Letters to Bishop during World War II.* Personal accounts of activity from the battlefield during World War II, 1942-45.

Curtis, Dr. Robert H. Medicine: *Great Lives.* New York: Atheneum, 1993.

Dale County Stories. *A History of Fort Rucker, Alabama.* Key word: Camp Rucker. [Online] http://www.graceba.net/~library/Dale.County.Stories /a.fort.rucker.html, 2003.

Dunnigan, James F., and Nofi, Albert A. *Dirty Little Secrets of World War II.* New York: Quill William Morrow, 1994.

Gilbert, Martin. *The Second World War* (First Owl Book Revised ed.) New York: Henry Holt and Company, 1991.

Global Security. Vietnam War. Keyword: *Vietnam War.* [Online] http:// www.globalsecurity.org/military/ops/vietnam.htm, 1976.

Greenleaf, Robert L. *87th Chemical Mortar Battalion, U.S. Army.* Key word: 87th Chemical. [Online] http://www.4point2.org/hist-87A.htm, 1945.

Heaton, Lt. Gen. Leonard D. *Medical Department, U.S. Army Surgery in World War II: Activities of Surgical Consultants, Vol. II.* Washington: U.S. Government Printing Office, 1964.

Heron, Azelia. *Life Stories.* Taped interviews. Hopedale, 1999.

Heron, Carol, and Heron, Debbie, and Heron, Larry, Jr., and Heron, Patty (1999) *Anecdote's.* Interviews. Hopedale, 1999.

Heron, Lawrence J. *Personal Scrapbook.* Sports and World War II clippings. Hopedale, 1935-39.

History 87th Chemical Mortar Battalion, Motorized. Battalion Daily Journal, 1943-45.

History of Chemical Disarmament. History. Key word: History of Chemical Disarmament. [Online] http://www.opcw.org/basic_facts/html/bf_int_main_frame_history.html, 2003.

History of Hq. & Hq. Company, 87th Chemical Bn. Company Daily Journal, 1943-44.

Holts, Major, and Holts, Mrs. *Battlefield Guide to the Normandy Landings.* South Yorkshire: Leo Cooper, 1999.

Kortegaard Engineering. *History of the 4.2 Chemical Mortar.* Key word: 4.2-inch mortar. [Online] http://www.rt66.com/~korteng/SmallArms/4pt2.htm, 1969.

Miller, Russell. *Nothing Less Than Victory.* New York: Quill, 1993.

Pravos, Edward F. Neptunus Rex: *Naval Invasion.* June 6, 1944. Novato: Presidio, 1998.

Ramsey, Gordon Clark. *Aspiration and Perseverance: The History of Avon*

Old Farms School. Avon. The Avon Old Farms School, Inc, 1984.

Roosevelt, Jr., Gen. Teddy. *Love Letters to his wife, Bunny.* Keyword:
Famous Love Letters. [Online] http://www.theromantic.com/LoveLetters/
rooseveltjr.htm, 1943.

Ryan, Cornelius. *The Longest Day.* New York: Touchstone, 1994.

Schmitt, Hans A. *Treaty of Versailles: Mirror of Europe's Postwar Agony.*
Key word: Treaty of Versailles. [Online] http://www.nv.cc.va.us/home/
cevans/ Versailles/papers/Schmitt_paper.html, 1989.

Scripps-Howard. *Reporting World War II.* New York: The Library of
America, 1944.

Spann, Edward K. *Hopedale: From Commune to Company Town 1840 –
1920.* Ann Arbor: Braun-Brumfield, Inc, 1992.

University of Notre Dame. Knute Rockne's *Win One for the Gipper.* Speech. Key word: The Gipper. [Online] http://www.google.com/
search?hl=en&ie=UTF-8&oe=UTF-8&q=the+gipper&btnG=Google+Sea
rch, 1940.

US Seventh Army Report of Operations. *Approach to the Siegfried Line.*
Key word: Forbach. [Online] http://www.trailblazersww2.org/approach_
to_the_siegfried_line.htm, 1988.

ABOUT THE AUTHOR

Greg Page authored best seller Camel Red, the story of Larry Heron's courageous journey to the shores of France during the Normandy Invasion, where he suffered a fate worse than death. Flooded with requests from readers to include photographs, Greg re-edited Camel Red, renaming it Beyond Recognition, and added 52 photographs to further illustrate Heron's painful ordeal. Look for Greg Page's next book, The Domino Conspiracy, which represents somewhat of a departure from the biographical novel. Although based on absolute truths and scientific facts, The Domino Conspiracy is a work of fiction that's been described as an explode-off-the-pages compelling thriller. A career officer in the U.S. Army, following graduation from West Point, Greg began a second career in the computer world, a perfect compliment since computers play a central role in every book project he now has in progress. An avid reader and historian, Greg collected rare books and created an extensive World War II library during his more than a decade of research for Beyond Recognition. Worldly experience and years of travel add a solid base from which to launch new fiction projects, aided by a powerful imagination and a thirst that only putting words to paper can allay.

www.ingramcontent.com/pod-product-compliance
Lightning Source LLC
Chambersburg PA
CBHW031944090426
42739CB00006B/80